Science Brain-Twisters, Paradoxes, and Fallacies

Science Brain-Twisters, Paradoxes, and Fallacies

Christopher P. Jargocki

Diagrams by Richard Liu

Charles Scribner's Sons / New York

To my mother

Copyright © 1976 Christopher P. Jargocki

Library of Congress Cataloging in Publication Data

Jargocki, Christopher P
  Science brain-twisters, paradoxes, and fallacies.

  1. Science—Problems, exercises, etc.
2. Engineering—Problems, exercises, etc.
I. Title.
Q182.J37   507'.6   75-31596
ISBN 0-684-14532-4

1 3 5 7 9 11 13 15 17 19 v/c 20 18 16 14 12 10 8 6 4 2

Printed in the United States of America

Contents

Preface

This book contains over 160 puzzles based on scientific principles, each with a detailed solution. The problems deal with such topics as space and time, mechanics, liquids and gases, automobiles, airplanes, science in sports, sound, heat, electricity and magnetism, radio, television, light, weather, geophysics, astronomy, and space science.

As the title indicates, the puzzles are of three kinds. There are brain-twisters: a bucket of water is placed on a scale. Will the reading change if you dip a finger in the water without touching the bucket?, paradoxes: Why do professional racing drivers accelerate when going around a curve?, and fallacies: Which is heavier: dry air or humid air?

As these examples would show, most of the puzzles contain an element of surprise. Indeed, the clash between common-sense conjecture and scientific reality is the central theme that runs through the book.

The puzzles range in difficulty from a few simple tongue-in-cheek questions to subtle problems requiring a lot of thought. Most puzzles are non-mathematical and require only a qualitative application of broad physical principles. All are designed to reward the reader with a great deal of physical intuition and insight into the world around him. If one can understand these knotty problems, the rest of classical physics should become much easier.

Preface

The book can be read with profit by high school and college students, science teachers, and all puzzle enthusiasts.

Several of the problems have been drawn from outside sources, which are acknowledged as follows: 1, D. W. Tomer; 21, attributed to Lewis Carroll; 28, Richard M. Sutton; 32, Sir Arthur Schuster.

Many people have helped me in the writing of this book. Special thanks are due to the late C. L. Stong of *Scientific American* who in his capacity as advisory editor helped me bring the manuscript down to a manageable size and made numerous valuable suggestions and criticisms. I am also grateful to my mother, Stefania Vcala, for encouragement and help in hunting down certain invaluable references; to Kristina Gubanski and Mark Pilate for assistance in researching the book, and to Susan Isaacs for typing most of the manuscript. Last but not least I want to offer my greatest thanks to my wife, Krystyna, who gave me emotional and financial support, thus creating an environment in which I could devote myself full time to writing.

Science Brain-Twisters, Paradoxes, and Fallacies

1. Space-Time Odyssey

1.

Imagine a sphere of diameter d. Its surface-to-volume ratio is

$$\frac{\pi d^2}{1/6 \pi d^3} = \frac{6}{d}$$

Now imagine a cube whose edge is also d. Its surface-to-volume ratio is

$$\frac{6d^2}{d^3} = \frac{6}{d}$$

The cube and sphere appear to have the same surface-to-volume ratio; but the sphere is said to have the smallest surface-to-volume ratio of any solid. Something is amiss. What is it?

2.

Two spherical mercury droplets combine into one droplet. It is warmer than the original two. Why?

3.

[a]   [b]

The diagram shows two leaves of the same shape.
The distance between tip and stem in (b) is three
times as large as the corresponding distance in (a).
Assuming for convenience that the leaves are flat, can
you say anything about their relative areas?

4.

You decide to build a fence enclosing the greatest
possible rectangular area, using 20 yd of fencing.
What should be the ratio of length to width?

5.

The gravitational force between two point masses is given by the inverse-square law

$$F = G \frac{m_1 m_2}{r^2}$$

Inverse-square laws also govern the force between two electric charges at rest, as well as the intensity of light and sound. What makes the inverse-square law so general in application?

6.

What was the first day of the 20th century?

7.

A man is celebrating his 29th birthday. How old is he?

8.

It is late June. A group of boy scouts, on a hike somewhere in New York State, cannot find their way back to camp. A debate starts on whether to stop for the night or keep groping in the dark. The decision depends on what time it is, but no one has a watch. Fortunately one of the boys is an amateur astronomer. One look at the first-quarter moon tells him that it is about two-thirds of the way down toward the horizon. He also remembers that the sun sets in this area at

5

about 7:30 EST in late June. On the basis of this information he quickly calculates the time.

How does he do it and what result does he get?

9.

Mr. X, being very tired, went to bed at 9:00 p.m. after setting his alarm clock for noon. When the alarm woke him up, how many hours had he slept?

10.

Why do ruled sandglasses have a tapering, "hourglass" shape?

11.

If you take your watch to the mountains, will it run fast or slow?

2. Motion

12.

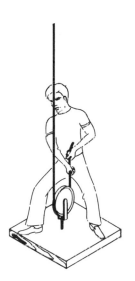

Can the man in the illustration lift himself and the block off the ground?

13.

Is a baby carriage with 2 ft wheels easier to push than one with 1 ft wheels?

14.

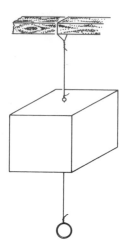

The drawing shows a heavy body suspended by a slender thread. A light handle is suspended from the body by a similar thread. What will happen if (1) you pull the handle down slowly, (2) you suddenly jerk the handle down?

15.

Bricks are stacked so that each brick projects as far as possible over the brick below without falling. Can the top brick project more than its length beyond the end of the bottom brick?

16.

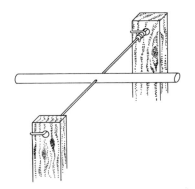

A steel rod is free to rotate about a thin steel string which passes through a hole drilled exactly through the center of the rod, as illustrated. In what position will the rod stop if you give it a push?

17.

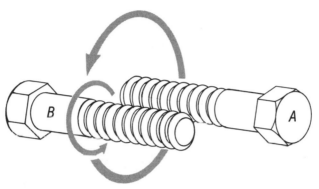

The illustration shows two identical bolts held together with their threads in mesh. While holding bolt A stationary you swing bolt B around it. (Don't let the bolts turn in your fingers.) Will the bolt heads get nearer, move farther apart, or remain at the same distance?

18.

Which is easier: pushing or pulling a wheelbarrow?

19.

What's wrong with this proof that $1 = 2$? Elementary mechanics includes the formula

$$v = at, \text{ or } a = \frac{v}{t}$$

where v is the speed, a is acceleration, and t is time. In another well-known motion formula,

$$s = \tfrac{1}{2}at^2, \text{ or } a = \frac{2s}{t^2}$$

where s is distance. Pairing the two equivalents of a we have

$$\frac{v}{t} = \frac{2s}{t^2}$$

Now multiply both sides by t:

$$v = \frac{2s}{t}$$

But $s/t = v$ (definition of velocity). Then

$$v = 2\left(\frac{s}{t}\right) = 2v$$

and

$$1 = 2$$

20.

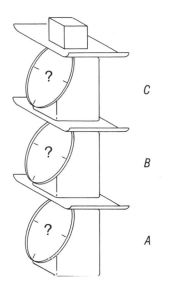

Three spring scales (A, B, and C), each weighing 2 lb, are placed on top of each other as shown. A 30 lb object is placed on the tray of the top scale. The object presses on all three scales. What weight does each scale indicate?

21.

A long rope passes over a pulley. A bunch of bananas is tied to one end of the rope, and a monkey of the same mass holds the other end. What will happen to the bananas if the monkey starts climbing up the rope?

Ignore the weight of the pulley and rope, as well as the friction between them.

22.

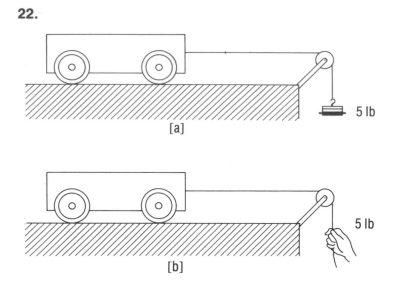

The two carts shown have equal masses and are accelerated by identical 5 lb forces, but the cart in (a) is accelerating slower than the cart in (b). This seems to contradict Newton's second law, which states that equal forces give equal accelerations to equal masses.

What is the way out of this paradox?

23.

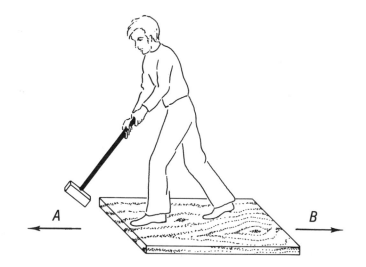

A man stands on a wooden plank and hits one end of it with a heavy sledgehammer (see illustration). You very likely did something of the sort as a child and found you could propel yourself along the floor.

Doesn't this violate Newton's first law? A body remains at rest or in a state of uniform motion unless acted upon by an *external* force. The sliding friction between the plank and the floor is the only relevant external force—unfortunately, sliding friction acts opposite to the motion of the plank, so how can it propel the plank forward?

Visualize the man and the back of the plank covered by a big box that gives the man enough room to swing the hammer. The box on the plank jerks forward with no apparent outside help.

24.

Even if you stand perfectly still on an accurate scale, the reading keeps oscillating around your average weight. Why?

25.

Suppose you throw a stone straight up so that it takes 3 seconds to reach maximum height. *Considering air resistance*, how long does it take the stone to fall to its starting point: less than 3 seconds, 3 seconds, or more than 3 seconds?

26.

Take two identical stones. Drop one from a given height above the ground, and simultaneously throw the other horizontally as far as you can from the same height. Which stone will reach the ground first, (1) neglecting air resistance, (2) considering air resistance?

27.

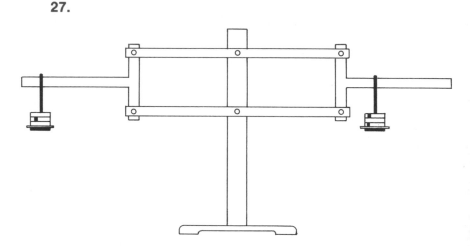

The two equal weights shown are free to slide on horizontal bars attached to a sort of pantograph constructed so the links always remain vertical and the longer bars always remain parallel as the system tilts one way or the other. The left-hand weight has just been moved out farther than the right-hand weight. Which end, if any, will go down?

28.

Here are two simple experiments to try. (1) Take a yardstick and a heavy object such as a rock or paperweight. Place the object on the right end of the stick and hold the stick horizontally, using your index fingers. Release both ends at the same time. You will see the stick and the object fall together.

(2) Repeat the experiment, but this time place the index finger of your left hand under one end of the stick. Release the right end of the stick so that it will be forced to rotate about your left index finger while falling. Now you will see the stick fall faster than the object.

Since the object is falling with the acceleration of gravity, the stick's acceleration must be greater than g. How can this be?

3. Liquids and Gases

29.

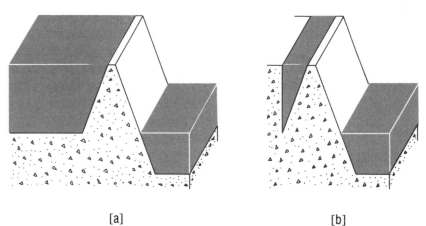

[a]   [b]

The diagram shows two dammed-up reservoirs of the same depth and width, both filled with water. One is, say, a mile long, and the other is very short, but both have the same end triangular cross section. The dam in (a) contains a huge volume of water; the dam in (b), only a tiny volume of water, comparatively. Does dam (a) have to be stronger than dam (b)?

30.

The jar in the picture has been filled with water and stopped up, using a rubber stopper with a straw passing through a hole. Can water be sucked out of the jar through the straw?

31.

A bucket of water is placed on one pan of a balance and an equal weight on the other. Will the equilibrium be disturbed if you dip a finger in the water without touching the bucket?

32.

What is wrong with this proof that hydrogen is identical with chlorine?

$$H_2 + Cl_2 = 2\,HCl$$

But this is the same as

$$HH + ClCl = 2\,HCl$$

Transposing, we get

$$HH - 2\,HCl + ClCl = 0$$

Then factor:

$$(H - Cl)^2 = 0$$

Hence

$$H - Cl = 0$$

or

$$H = Cl$$

33.

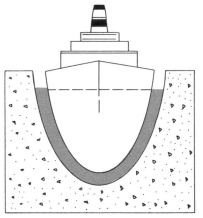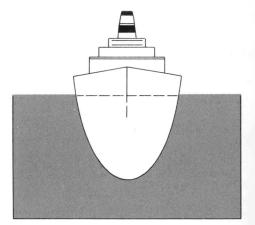

[b]   [a]

In the figure, (a) shows a liner on the high seas. It weighs, say, 50,000 tons. Imagine (b) the same ship slowly lowered into a dry dock shaped like the ship but a little larger and filled with water. As the ship is lowered the water is forced out of the dock until there is only a thin water envelope between the hull and the dock walls.

Will the ship float on the thin layer of water or will it touch bottom and squeeze the rest of the water out?

34.

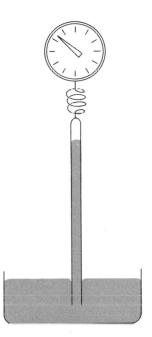

A barometer tube with mercury in it is suspended from a spring scale, as shown. Will the scale show the weight of the tube or the weight of the tube plus the mercury in it?

19

35.

Can a dry plastic sponge soak up more water than its volume?

36.

When a submarine sinks to a clay or sandy bottom, it is sometimes unable to lift itself off, as if glued to the bottom. Why?

37.

Which weighs more—a pound of feathers or a pound of iron? ("They weigh the same" is not a permitted answer.)

38.

An aquarium has two compartments, one very small and one large, formed by a vertical membrane made of thin rubber (see drawing). The water in the small compartment is at a higher level than in the large one. Which way, if any, will the membrane bulge?

39.

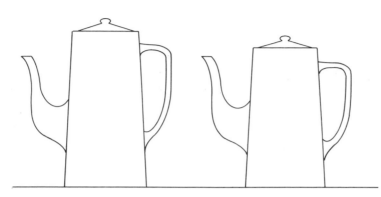

We show two coffeepots having the same cross-sectional area. The first is taller than the second. Which of the two (if either) will hold more coffee?

40.

A block of wood floats in a glass of water. The glass is placed in an elevator. Will the block stick out higher above the surface of the water when the elevator starts down with an acceleration $a < g$?

41.

A balloon is seen on a hot day traveling horizontally well overhead. Do the occupants of the balloon feel any wind?

42.

Place a pail filled to the brim with water on one pan

of a scale. On the other pan, place another pail of water filled to the brim—except that it has a piece of wood floating in it. Is the second pail lighter than the first?

43.

Inside a moving automobile a child holds a helium balloon by a string. All the windows are closed. Which way will the balloon move if the car makes a right turn?

44.

You are in a car, waiting in the left-turn lane. A stream of cars is passing you on the right at high speeds. Which way will your car sway because of the passing cars: to the right, to the left, or not at all?

45.

Which is heavier: humid air or dry air?

46.

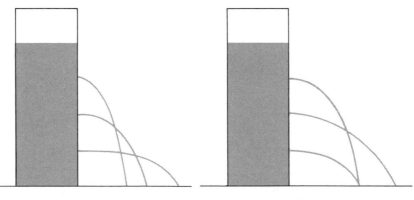

[a]   [b]

A water can has three holes spaced at equal intervals, the middle hole being halfway up the column of water. The figure shows two ways the water might flow out of the holes. Which one is correct?

47.

An arrow is pointed at an angle above the horizontal and shot from a bow. As it flies, the arrow somehow manages to rotate itself so that it is always tangent to the parabola of flight, and eventually strikes the ground point first.

How is this possible?

48.

Watching rafts floating down a river, we see that those close to the center float faster than those near the banks. Also, heavily loaded rafts float faster than lightly loaded ones. Why?

49.

This problem is generally known as Dubuat's paradox.

Suppose you hold a stick in a stream flowing with speed v. Then you tow the same stick with speed v through still water. All motion is relative, and it makes no difference which is moving—water or stick—as long as their relative velocity is the same.

One might conclude that water resistance would be the same in both cases. Is it?

4. Earth Travel

50.

Automotive engineers often say the front wheels of a conventional car are better stoppers than the rear wheels. On many cars the front brakes are bigger and stronger or are disk brakes, which, being exposed to the airstream, do not overheat as easily as drum brakes, and thus hold on to more of their braking power.

Why the reliance on front wheels?

51.

You are driving fast on a highway when suddenly a large boulder rolls off a dump truck and stops in your lane, 400 ft ahead. You stab the brakes hard and your car is thrown into a long, screaming skid. As the car closes in on the boulder, you while away the time looking for a theory to explain why the front of your car goes into a nose dive when you apply the brakes. Any ideas?

52.

In our childhood most of us played with model cars. Imagine that you have two identical model cars, one black and one white. You lock the front wheels of the white car and the rear wheels of the black car—say, by putting a piece of paper between the wheels and the body. Then you release the cars at the top of a slippery board. Can you predict what will happen? Will one car, or both cars, descend with its front forward?

53.

Car owner's manuals advise us to use the car's engine as a "fifth brake" when descending a long steep grade. They say the engine braking effect is greater in second than in high gear, and is greatest in low gear. Why is this?

54.

A car traveling at 70 mph on a level road is placed in neutral gear without braking. Despite air resistance, engine braking effects, and the inertia of the drive-line, at the end of a mile the car is still going about 15 mph. This illustrates how small rolling friction is compared to sliding friction. Why is it so small?

55.

Steam locomotives and electric cars do not need transmissions, but cars powered by internal-combustion engines do. Why?

56.

It is strange but true that professional racing drivers accelerate when going around a curve. Why?

57.

The purpose of the tread on tires is to increase their grip on the road. If you agree with this statement, then (1) why do drag racers use "slicks," that is, tires with no tread on them, (2) why do brake linings have no tread?

58.

Imagine that identical cars A and B collide and:
    1. Car A is traveling at 30 mph, as is car B;
    2. Car A is traveling at 50 mph, car B at 10 mph;
    3. Car A is traveling at 60 mph, car B is stationary.
    The relative velocity of the cars is always the same—60 mph. Will the damage in each accident be the same?

59.

A car is moving north at high speed. At a highway exit the driver, without slowing down, makes a sharp turn to the east. If a pair of wheels leaves the roadway, will it be the pair on the inside of the turn or the pair on the outside?

60.

Mr. X is driving fast. A strong wind is blowing from the left, but fortunately the road is dry, so the car has no problem staying in its lane. Suddenly the driver ahead of Mr. X slows down, forcing him to step on the brakes. Mr. X applies the brakes too hard, the wheels lock and slide over the highway. Unexpectedly, the wind now easily pushes the car into the next right lane, as if the road had turned into slippery ice.

Why can't a car sliding forward resist a lateral force like the wind?

61.

You are driving fast along a side road which ends in a T-shaped intersection with a highway. There is a wall along the far side of the highway, and no car is visible in either direction. What should you do to avoid hitting the wall—steer straight at it, fully applying the brakes, or turn left in a circular arc as you enter the highway, using all the available force of friction to produce centripetal acceleration?

62.

A car drives the first lap of two at 30 mph. How fast must it drive the second lap to average 60 mph for both laps?

63.

Conestoga Wagon

Front wheel of modern bicycle

Compare the wheels in the figure. On the modern bicycle the spokes are mounted tangentially; on the Conestoga wagon they are mounted radially. Why?

64.

In horse-and-buggy days artists often gave the impression of moving wheels by showing distinct spokes below the axles and blurry ones above them. Does the top of a rolling wheel actually move faster than the bottom?

65.

Why is it so easy to go into a skid when descending a slippery grade?

5. Armchair Athletics

66.

If you pick up a heavy suitcase and hold it off the ground for a while you will begin to sweat, shake, and breathe harder than if you were running up a flight of stairs. Yet if work equals force times displacement, in the physical sense you are doing no work whatever. You might as well be replaced by a table. The latter will hold up the suitcase as long as desired with no effort and no external source of energy. By contrast, humans do need an external source of energy, that is, food, to do the same thing, so they must be doing some kind of work.

Can you resolve this paradox?

67.

Pound for pound of lean meat, women are as strong as men. True or false?

68.

See if you can do this stunt. Face the edge of an open door with your nose and stomach touching the edge and your feet extending forward slightly beyond it. Now try to rise on tiptoe.

Why is it impossible?

69.

The grasshopper can leap about 10 times its body length in a vertical jump or 20 times its length (almost 1 m) horizontally. Cat fleas and human fleas can jump to a height of 33 cm (over a foot)—a hundred times their own length!—developing an acceleration of 140 g's. If a man could do that well in proportion to his height, he could jump over a 50-story building.

Why can't he?

70.

Challenge somebody to do this trick. He is to stand straight with his back and his feet touching a wall. Now ask him to bend over and touch his toes without bending his knees. Even if he is in good shape he will not be able to do it without falling over. Why not?

71.

In Oslo a champion high jumper clears the bar at 7 ft 5 in. In Mexico City, a second champion wins with a jump of 7 ft 5⅛ in. Who has made the better jump?

72.

Why is it necessary to use complicated scuba equipment or bulky diving suits? Why not just breathe through a long hose or snorkel whose upper end is attached to a float?

73.

Is it possible for a high jumper's center of gravity to pass under the bar while he passes over it?

74.

A child's swing is often used to illustrate the phenomenon of resonance. If you give a child a push at the top of each swing, almost all the energy of the push will go into increasing the child's kinetic energy. But the child, with some practice, can achieve the same result without any outside help, by what is known as "pumping." What is the physics of pumping?

75.

Do baseballs really curve, and if so, how?

76.

Many baseball players and spectators insist that they see a pitched ball travel in a straight line, then curve suddenly just before it reaches home plate. Can a pitched ball change direction suddenly?

77.

Why are golf balls dimpled?

6. The Realm
of Flight

78.

Airflow

Airflow

[a]   [b]

In (a) of the drawing an airfoil is placed in an air-stream moving to the right at, say, 200 mph, with its knifelike edge facing upstream. In (b) the same airfoil has been turned around and placed with its rounded edge facing upstream.

In which position will the airfoil offer less resistance to the air?

79.

A plane is flying along the perimeter of an equilateral triangle ABC. There is no wind, and the plane is flying at constant speed. The trip from A to B takes 1 hour 20 minutes, and the trip from B to C also takes 1 hour 20 minutes. How come the trip from C to A only takes 80 minutes?

80.

[a]   [b]

The drawing shows an airfoil 10 in. thick and a round wire 1 in. in diameter. Which shape will produce less drag?

81.

Some helicopters have two propellers on vertical axes but rotating in opposite directions; others have one propeller on a vertical axis and the other at the tail on a horizontal axis perpendicular to the fuselage. Why don't helicopters have just one propeller?

82.

A plane flies with the wind from *A* to *B*, and against the wind back to *A*. The pilot saves as much time flying with the wind as he loses flying against it, so the round trip will take as long as if there were no wind. Do you agree?

83.

Why do airplanes generally take off into the wind?

84.

Air bumps are regions where the air has a higher than normal density; air pockets are regions with air of lower density. True or false?

85.

A jet flight from San Francisco to New York takes about 5 hours; a flight from New York to San Francisco takes about 6 hours. (Ignore the effects of changing time zones.) Why the difference?

86.

When an airplane rises above a layer of clouds the flying is no longer bumpy (with the exception of clear-air turbulence). Why?

7. Sounds and Voices

87.

Is it possible for a person or an animal to make sounds he (it) cannot hear?

88.

Most of us, on listening to a tape recording of our own voice, will swear that it does not sound like us at all. Are we victims of an illusion or is the difference real?

89.

Why do tuning forks have two prongs?

90.

A grasshopper emits sounds that can be heard for ½ mi. Given that the density of the air is 1.293 kg/m³, the mass of the air in a hemisphere ½ mi in radius is about 1 million tons. How can a tiny grasshopper move such an enormous mass of air merely by jiggling a part of its anatomy?

91.

How can a sound wave traveling down a tube get reflected from its open end, that is, from nothing?

92.

1. Why does sound travel unusually well on a calm, clear night?

2. Why does sound carry well over water, especially in summer?

3. How is it that mountain climbers and balloonists often hear and understand persons on the ground, even from a half mile up, while the latter cannot hear and understand them at all?

4. On February 2, 1901, cannons were fired in London to mourn the death of Queen Victoria. The sound was heard throughout the city but not in the surrounding countryside. Strangely enough, the cannon fire was clearly heard by astonished villagers 90 mi away. How could the sound hop over the outskirts of London and come down 90 mi away?

93.

Why is it difficult to hear upwind from a source of sound, apart from the masking effect of the noise produced by the wind? Is it because the wind "blows" the sound back?

94.

Why are loudspeakers mounted on solid boards with only a hole for the mouth of the speaker—and sometimes completely enclosed from the rear?

95.

People who live in a cold climate often observe how quiet it gets during and after a considerable fall of fluffy snow. Can you explain this?

96.

A north-south freeway is proposed to run outside a city. Considering the health of the inhabitants as a major factor, should the freeway run west or east of the city?

97.

If a tuning fork held close to one ear is slowly rotated about the axis passing through the handle, you will hear the sound grow louder and softer.

You may think the cause is mutual interference of the sound waves produced by the two prongs. However, if the fork vibrates at 440 Hz, the wavelength produced is

$$\lambda = \frac{340 \text{ m/sec}}{440 \text{ sec}^{-1}} = 0.77 \text{ m}$$

For destructive interference to occur, the difference in path lengths from the prongs to the ear must be ½ wavelength, that is, about 40 cm. Since the distance between the prongs is only 2–3 cm, interference cannot account for the wavering of the sound. What does account for it?

98.

Why does thunder roll?

8. Heat

99.

Given three identical Dewar flasks *A*, *B*, and *C*. (Thermos bottles are Dewar flasks.) There is also an empty container *D* which has thermally perfectly conducting walls and which can be easily placed inside a Dewar flask.

Pour 1 liter of water into *A* and 1 liter into *B*, water temperatures being 80° C and 20° C respectively. Now, using all four containers, can you heat the cold water with the aid of the hot water so that the final temperature of the cold water will be higher than the final temperature of the hot water? (You may not mix the hot with the cold water.)

100.

The boiling point of water is lower when atmospheric pressure decreases. Therefore, why not pump out some of the air above the water in a kitchen pot, obtaining boiling water faster and saving some energy?

101.

Normal human body temperature is about 37° C (98.6° F); it varies by about 0.5° C depending on the time of day, reaching a maximum at 4:00–5:00 p.m.

Room temperature is about 18°–22° C, which is 15°–19° C lower than body temperature. Shouldn't we be constantly shivering to make up for the enormous losses of heat by radiation?

102.

Why is it hard to ice-skate when it's very cold?

103.

If you take your coffee with cream, here is a problem for you. You are in a hurry to catch the 7:25 bus. What should you do to make the morning coffee cool off faster: pour the cream, which is cold, in it first and wait 5 minutes before drinking, or wait 5 minutes before adding the cream?

104.

When you pour hot water into a thin glass, is it more likely, as likely, or less likely to break than a thick glass?

105.

Why are the freezing compartments of the most efficient refrigerators placed at the top?

106.

Can ice be colder than 0° C (32° F)?

107.

The experts who design clothes to keep people warm in polar climates say you should wear a hat there. Why?

108.

Is there a difference between a gas and a vapor?

109.

When the engine overheats, some drivers stop and take off the radiator cap without waiting for the engine to cool off. If there is any water left, it usually spouts like Old Faithful. Why?

110.

Drinking a cup of iced tea on a hot day will certainly cool you off. Can you cool off at all by drinking a cup of hot tea?

111.

Why does snow squeak under your shoes on a very

cold day, but not when the temperature is just below freezing?

112.

Generally speaking, small animals are more vulnerable to cold than large animals. Why?

113.

If you put a spoonful of coffee grains in water at slightly below 100° C, the water seems to suddenly start boiling. What's going on?

114.

Try this parlor trick on your friends: Float an ice cube in a glass of water, and challenge someone to remove the ice cube using only a paper match as a tool. When he or she gives up, bend the head of the match to a right angle, place the body of the match flat on top of the ice, and sprinkle it with salt until completely covered. The match will quickly freeze to the cube and you will be able to lift the cube out of the glass by grabbing the head of the match.
    Explain the trick.

115.

Ice cubes in a bucket have an annoying tendency to stick together. Why?

116.

A gallon of cold gasoline gives more mileage than a gallon of warm gasoline. True or false?

117.

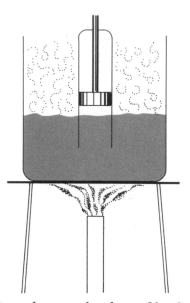

The illustration shows a beaker of boiling water; immersed in the water is a glass tube containing a movable piston. Suppose you start with the piston touching the surface of the water. If you slowly raise the piston, the boiling water will *not* move up—although at room temperature it would, because of atmospheric pressure. Why the change?

118.

The density of water increases with lowering temperature down to 3.98° C, reaching a maximum value of 1.00000 g/ml. So far water behaves like most substances; but between 3.98° and 0° C its density decreases to 0.99987 g/ml. Upon freezing, its density decreases further to 0.9168 g/ml. Thus in turning into ice, water expands by about 11 percent, exerting a force strong enough to burst water pipes. Is water the only substance that behaves this way?

119.

Why is it that you can warm your hands by blowing gently and cool them by blowing hard?

9. Electricity and Magnetism

120.

When you step outdoors on a clear day you are surrounded by a vertical electric field of 100–500 volts per meter. The field is directed downward and is set up by the positive charge in the atmosphere. When a charged thundercloud comes along, the field may go up to 10,000 volts per meter. Why doesn't this voltage kill you?

121.

Two people accidentally touch a live wire carrying a 110 volt alternating current. One of them dies; the other only receives a mild shock. How can this be?

122.

Suppose you pass a current through a wire. If the current is strong enough the wire will get noticeably warmer. If you then cool one part of the wire, the other part will get much warmer than before cooling. Why?

Assume the same potential difference is maintained across the wire in both steps.

123.

The only difference between two steel bars is that one is a permanent magnet and the other is unmagnetized. Without using any equipment, can you tell which is which?

124.

An American visitor to the United Kingdom may be astonished to see TV antennas on rooftops with vertical elements instead of horizontal ones as in the United States. Why the difference?

125.

An electric current flowing through a conductor induces a magnetic field around it. The drift velocity of electrons making up the current is only a few millimeters per second. If an observer walks along the conductor in the direction of and at a speed equal to the drift velocity of the electrons, relative to him the electrons will become charges at rest. Will the magnetic field around the wire also disappear for the walking observer?

126.

Here is a simple experiment. With the TV on, take a ruler or stick and wave it rapidly in front of the screen. You will see many "frozen" rulers radiating from your hand. The experiment works even better if you cover the screen except for a horizontal strip a few inches wide. Evidently, a TV set can be used as a stroboscope. Why?

(A fluorescent lamp also makes a good stroboscope.)

127.

Why is shortwave reception better at night?

128.

Why have most countries switched from 110 to 220 or 240 volt household power levels?

129.

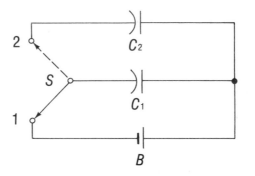

The diagram shows a circuit with two capacitors $C_1 = C_2 = 10\mu f$, a battery B with an electromotive force of 20 volts, and a switch S. Suppose that the switch is initially in position 1. Capacitor C_1 will be charged up by the battery, and the energy stored in it will be

$$W_1 = \frac{C_1E^2}{2} = \frac{10\mu f\,(20V)^2}{2} = 2 \times 10^{-3}\,\text{joule}$$

If you throw the switch to position 2, the capacitors are connected in parallel, and the charge originally on C_1 will have a chance to arrange itself on both C_1 and C_2. The combined capacitance of the two capacitors is $10\mu f + 10\mu f = 20\mu f$. However, since the original charge has divided itself equally between C_1 and C_2, the potential difference across the terminals of this "compound capacitor" is only 10 volts. Thus, the energy stored in the compound capacitor will be

$$W_2 = \frac{20\mu f\,(10V)^2}{2} = 10^{-3}\,\text{joule}$$

Half of the energy originally stored in C_1 is missing. What happened to it?

130.

If a magnetic needle is placed on the surface of water, it positions itself along the magnetic meridian (that is, a great circle joining the magnetic poles of the earth), but does not move north or south. But if the same needle is placed near a strong magnet, it will not only rotate but will start moving toward it.

How do you explain this difference?

131.

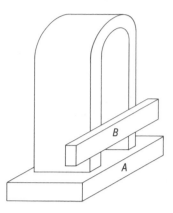

We close a strong horseshoe magnet with an iron bar
A (see diagram). The magnet is strong enough to hold
it in place. Then we take a bar B, made of soft iron,
and place it on the magnet as shown. As soon as
we do, bar A drops off. When bar B is removed, the
magnet easily holds bar A up again.

How do you explain this phenomenon?

132.

[a]

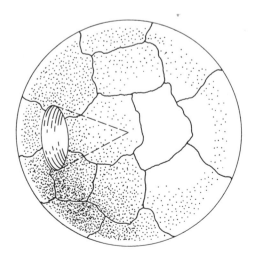

[b]

In school we learn that a magnet has two poles. If someone could build a magnet with *one* pole, the next Nobel prize in physics would be his. Here's one modest proposal.

Cut a steel sphere into irregular sections shaped as in the diagram (a). Magnetize each section so that all the round ends become north poles and all the sharp ends south poles (or vice versa). Put the magnetized sections together so they again form a sphere (b). Will the magnetized sphere have the north pole on the outside? Has the south pole disappeared?

10. Light and Vision

133.

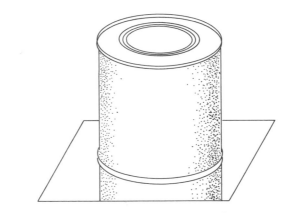

The diagram shows a metal can standing on a mirror. Is the mirror made of glass or polished metal?

134.

A narrow beam of white light goes through a glass prism that disperses the beam into its constituent colors. Can the colors be recombined into white light by transmitting them through an identical, inverted prism?

135.

If H. G. Wells's invisible man existed, could he see?

136.

At sunset or sunrise it is sometimes possible to see the "green flash." Just as the last of the solar disk is about to disappear or the top of the disk to appear, for a fraction of a second it turns a brilliant green. The effect can only be seen if the air is clear and the horizon is distinctly visible—usually at sea or in mountain or desert country.
How does nature produce the green flash?

137.

The moon and planets look bigger and closer when viewed through a telescope. But the stars, even through the largest telescopes, appear as point sources of light. So what is the use of astronomical telescopes except for looking at planets?

138.

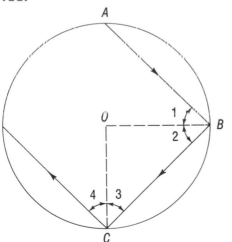

In explaining the rainbow it is often assumed that a light ray enters a raindrop at A (in the diagram), undergoes total internal reflection at B, and leaves the drop at C. At A and C the ray crosses the boundary between air and water, undergoing dispersive bending. The result is that the light leaving at C is split into every color of the visible spectrum, which produces the rainbow.

However, it is easy to show that a light ray that has undergone internal reflection once will never get out of the drop.

Assume that angle 1 in the figure is greater than the critical angle necessary for internal reflection. Then at B the light ray will undergo an internal reflection and will go toward C. Since triangle OBC is isosceles, angles 2 and 3 are equal to each other. The angle of incidence 1 is equal to the angle of reflection 2, and the angle of incidence 3 is equal to the angle of reflection 4. Hence all four angles are equal. Therefore, if angle 1 is greater than critical, so is angle 3. Then there will be another internal reflection at C, and so on, and the ray stays inside the drop forever.

How can we explain the formation of the rainbow?

139.

From a plane flying above the ocean the water looks much darker directly below than toward the horizon. Why?

140.

The wavelengths of light λ passing through two mediums are given by

$$\frac{\lambda_1}{\lambda_2} = \frac{v_1}{v_2} = n_{1,2}$$

where v_1 and v_2 are the speeds of light in the two mediums and $n_{1,2}$ is the refractive index of medium 2 relative to medium 1.

The wavelength of light changes as it passes from one medium into another. For example, if the wavelength is 0.65 micron in air (red), then in water, whose refractive index relative to air is 1.33, the wavelength changes to

$$\lambda_2 = \frac{\lambda_1}{n_{1,2}} = \frac{0.65\mu}{1.33} = 0.49\mu$$

which corresponds to blue light.

Then does a red lamp appear blue to a diver?

141.

Normally smoke looks bluish gray but when seen against the sun it turns reddish gray. What is its true color?

142.

Is there a way to tell a real landscape from one reflected in the water, on a photograph?

143.

Why is it that when you look at a neon sign red letters seem closer than blue or green ones?

144.

Why does wet sand look darker than dry sand?

145.

Suppose it is daytime and you are looking at the outside of a building. Why do the windows look darker than the walls, even if the walls are painted dark?

146.

Workers at open-hearth furnaces often wear protective clothing coated on the outside with a thin metal layer. This does not seem to make sense, since metals are excellent conductors of heat. Explain.

11. Spaceship Earth

147.

How does "pouring oil on troubled waters" really work? It is claimed that the sea surface is smoothed by increasing the surface tension of the nearby area of sea. But the surface tension of water is twice that of oil, so how does oil help?

148.

Why does the sea foam?

149.

In the U.S., Pacific Coast water is usually much colder than Atlantic Coast water—why?

150.

When a car is parked by a house wall overnight, in the morning the car windows near the wall may be dry while those on the far side are covered with dew. Explain.

151.

Why do grass and low-lying small plants get so wet at night, even in summer?

152.

Why does Antarctica have eight times as much ice as the Arctic?

153.

Unless small amounts of soda or magnesium carbonate, or the like, are added, ordinary salt will not pour in humid weather. Any explanation?

154.

How do smudge pots protect orchards from freezing?

155.

A tramp, walking down a road, comes to a place where five roads meet and finds the signpost lying in a ditch. No one is in sight, it is a cloudy day, and he has no compass. How can he tell which road goes to the town he is heading for?

156.

Where on earth will a compass needle point south with both ends?

157.

It snows, on unfrozen, moist soil. The grass will be covered by snow long after the snow on bare soil has melted. Why?

12. The Universe

158.

What is the effect of air drag on a satellite traveling through the upper layers of the atmosphere—to slow it down or speed it up?

159.

Why is it that space launch sites such as Cape Canaveral tend to be located toward the tropics?

160.

How can an astronaut pour a liquid from one container into another while weightless?

161.

When is the earth moving fastest around the sun? When is it moving the slowest?

162.

An open, clean glass jar containing water is taken aboard an orbiting spacecraft, where it becomes weightless. What happens to the water in the jar?

163.

A spaceship freely coasting around the earth is constantly turning. Why aren't the astronauts thrown against the wall as the occupants of a car are when the car rounds a curve at high speed?

164.

Why is it that most satellites can only be seen 1–2 hours after sunset or 1–2 hours before sunrise?

165.

An astronaut in a spacecraft puts a kettle of water on an electric stove to boil, under conditions of weightlessness. When he checks the kettle an hour later, the water on top is still cold. How come?

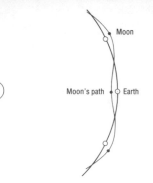

Moon

Moon's path • ○ Earth

○
Sun

166.

The diagram shows a section of the earth's orbit around the sun, with the moon's trajectory around the earth. Apart from not being drawn to scale, is there anything basically wrong with the diagram?

167.

If a body is more than 161,000 mi from the earth it is attracted more strongly by the sun than by the earth, as can be verified by calculation using the inverse-square law. But 161,000 mi is only two-thirds as much as the average distance from the earth to the moon; therefore, the moon is pulled more by the sun than by the earth—in fact, more than twice as much.

Then why doesn't the sun steal the moon from the earth?

168.

How can you tell a planet from a star with the naked eye, without waiting for the planet to move relative to the stars?

169.

A dying artificial satellite makes its final appearance at the same time and in the same part of the sky for several days before disintegrating. Why?

Answers

1. *Space-Time Odyssey*

1.

The sphere does have the smallest surface-to-volume ratio of any solid—for a given volume. Let a sphere have volume V. Then its diameter d, obtained from $V = \pi d^3/6$, is $(6V/\pi)^{1/3}$, and its surface-to-volume ratio is

$$\frac{\pi(6V/\pi)^{2/3}}{V} \approx \frac{4.8}{V^{1/3}}$$

A cube of the same volume V has edge $V^{1/3}$, and its surface-to-volume ratio is

$$\frac{6\,(V^{1/3})^2}{V} = \frac{6}{V^{1/3}}$$

Thus a sphere's surface area is 20 percent less than that of a cube of the same volume.

Also we can invert this reasoning to show that for a given surface area the sphere encloses a larger volume than the cube. The reader may verify that the difference is a whopping 39 percent.

Soap bubbles are spherical because they minimize the energy of surface tension by minimizing the area of their surfaces. Inversely, it can be shown that the largest possible enclosure with a given amount of wall material is spherical.

2.

The surface area of the new droplet is smaller than the total surface area of the original two droplets. Total volume has not changed, since the total amount of mercury remains the same. Decrease in area means decreased energy of surface tension that pulls mercury into spheres. The extra energy goes to heat the mercury in the droplet. (The temperature increase is very small—on the order of 0.01° C for 1 cm droplets.)

3.

The ratio of the irregular areas is exactly $3^2 = 9$. The ratio of the volumes of two similar solids, no matter how irregular, is n^3, where n is the ratio of the corresponding distances between *any* two points of the solids.

4.

The ratio should be 1:1, that is, the rectangle should be a square.

Rephrase the problem in algebraic terms: In what two parts should a number a be divided to make their product a maximum?

Let the two parts be $a/2 + x$ and $a/2 - x$, where x is any number between 0 and $a/2$. Their sum is

$$(a/2 + x) + (a/2 - x) = a$$

as required. Their product is

$$(a/2 + x)(a/2 - x) = a^2/4 - x^2$$

Now $-x^2$ is always negative or zero. The product is a maximum, $a^2/4$, when $x = 0$. Then a should be divided in two equal parts.

5.

A signal emitted by a point source spreads out in all directions, forming the surface of a sphere with radius r. The area of this surface in our three-dimensional space is $4\pi r^2$. The area of the sphere increases in proportion to the square of the distance between the surface and the center. Think of the strength of the signal as the inverse of this proportion.

6.

Most people will say January 1, 1900. But the first day of the 20th century was January 1, 1901.

The first century A.D. began on January 1, 1. On December 31, 99, there had been 99 years of the first century; to complete the first century we must add the entire year 100, ending on December 31, 100. Similarly, the 19th century ended on December 31, 1900.

One source of confusion is that many of us believe the Christian era started on January 1, 0. Then, as with people, January 1, 1, would mean "1 year old"; January 1, 100, would mean "100 years old";

and so on. The trouble with this is that there was no year 0. The year before 1 A.D. was 1 B.C.

7.

On his 29th birthday he is only 28 years old, that is, he has lived 28 full years. On your first birthday you had just been born, and it was only on your second birthday that you reached the age of one year. Thus the colloquial use of "birthday" with the number of one's age is literally incorrect.

8.

In the top figure we look down on the northern hemisphere: the earth is rotating counterclockwise on its axis. Similarly, the moon revolves counterclockwise around the earth. The phases of the moon as seen from the earth are also depicted.

To find the positions of the moon in the sky, pick a phase, say, the first quarter. This fixes the position of the moon. The position of the sun can be considered fixed. This leaves only the motion of the earth rotating on its axis. This is a fairly realistic approximation, for during 24 hours the earth covers only 1/365.26 of its path around the sun, and the moon, only 1/27.3 of its orbit around the earth (both relative to the "fixed" stars).

To show the rotation of the earth, pass a line through its center representing the local horizon (the line should be tangent to the earth, but for our purpose this makes no difference). As the line rotates counterclockwise through the noon, sunset, midnight, and sunrise positions (bottom figures) we see

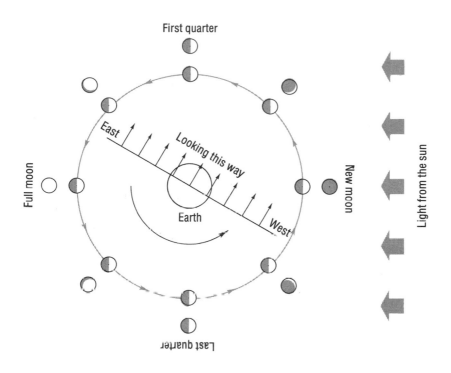

First quarter

East

Looking this way

Earth

West

Full moon

New moon

Light from the sun

Last quarter

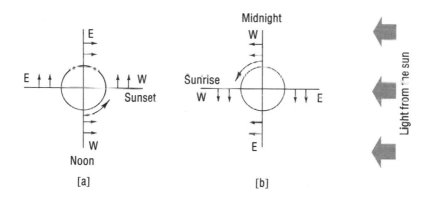

E

E ↑ ↑ ↑   ↑ ↑ W
Sunset

W

Noon

[a]

Midnight

W

Sunrise

W ↓ ↓ ↓   ↓ ↓ E

E

[b]

Light from the sun

69

(a) that the first-quarter moon rises at noon and reaches its highest position in the sky at sunset, and (b) sets at midnight. Repeating this procedure for the other phases, we obtain a table of the positions of the moon in the sky.

	Rises	Highest	Sets
New moon	Sunrise	Noon	Sunset
First quarter	Noon	Sunset	Midnight
Full moon	Sunset	Midnight	Sunrise
Last quarter	Midnight	Sunrise	Noon

(For the southern hemisphere, interchange the first and last quarters, since there the moon is "upside down.")

In the problem the moon is two-thirds of the way down from its highest position at sunset to its setting position at midnight. Sunset was at 7:30 p.m., so the time is two-thirds of the way from 7:30 p.m. to midnight, 10:30 p.m., *approximately.*

9.

Three hours. (If you said 15 hours, you should read problems more slowly.)

10.

To assure that the time scale on the glass is uniform, with equal distances between scale divisions corresponding to equal lengths of time. If the sandglass didn't taper, the top of the sand column would descend at increasing speed.

11.

Fast. The basic component of a watch is the balance wheel, which oscillates exactly 300 times a minute. The moment of inertia of the balance wheel is slightly higher in air than in a vacuum because of the viscosity of air—the balance wheel has to drag air around with it. In the mountains the density and viscosity of the air decrease slightly, allowing the balance wheel to oscillate faster.

12.

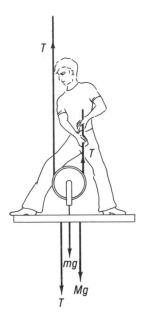

The man looks as though he is trying to lift himself up by his own bootstraps, which, Baron Munchausen's boastful stories notwithstanding, is impossible. But he is not. Actual tests have shown that a 190-pound man can lift not only himself this way, but a 110-pound block too.

In the diagram, the man (who weighs Mg) pulls up on the rope with a force T which produces a tension T in the rope. Considering the rope and pulley as weightless, and the pulley as frictionless, T is transmitted around the pulley and acts upward on it. By Newton's third law, in pulling up on the rope the man pushes down on the block with an equal force, in addition to his weight. The block (weight mg) is acted on by an upward force $T + T = 2T$, and by a downward force $mg + Mg + T$. If the block is in equilibrium the two forces must be equal, so $T = mg + Mg$. When the man pulls with a force greater than his own weight plus the weight of the block, he will rise from the ground.

13.

Yes, but only because the road is not perfectly smooth. Over a given distance the 1 ft wheel rotates twice as many times as the larger wheel, which involves more work done against friction.

Also, roads, besides being rough, may contain pebbles. The horizontal push necessary to force the 1 ft wheel over a pebble is greater than that needed for the 2 ft wheel. This is why the famous Conestoga wagons, used in the early westward expansion, had such large wheels.

14.

1. The upper thread will break. It has to support the weight of the body plus the weight of the handle and the downward pull on the handle; the lower thread has to support only the latter two.

2. The inertia of the heavy body will delay its movement while the force of the jerk builds up to a value that breaks the lower thread.

15.

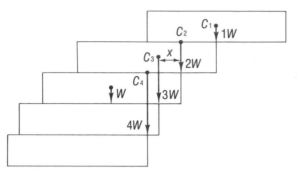

It can. And not only more, but as much more as we please.

In the diagram, if one brick is placed on another, the upper brick will not fall if its center of gravity is anywhere above the next lower brick. The greatest offset, equal to half a brick's length, is obtained when the center of gravity of the upper brick C_1 lies directly above the end of the brick below.

How can we place the two bricks on top of a third one to obtain maximum offset? The combined center of gravity of the two bricks, C_2, is one-fourth of a brick's length from the end of the second brick. We position C_2 over the end of the third brick. The second brick is offset one-fourth of a brick's length over the third brick.

To place these three bricks with maximum offset on a fourth one, we have to find their center of gravity C_3. The torque due to the combined weight of the upper two bricks about C_3 must equal the torque due to the weight of the bottom brick about C_3.

Let x be the arm of the total weight $2W$ of the two upper bricks, applied at C_2, relative to C_3. Then if L is the length of one brick, the arm of the weight W of the bottom brick relative to C_3 is $L - x$. We have the equation

$$W(\frac{L}{2} - x) = 2Wx$$

which gives $x = L/6$. The maximum offset of the third brick is one-sixth of a brick's length jutting out over the fourth brick.

Similarly, we derive the offset y of the fourth brick over the fifth. The equality of the torques about C_4, the center of gravity of the upper four bricks, gives the equation

$$W(\frac{L}{2} - y) = 3Wy$$

which gives $y = L/8$. It is evident that the equation giving the location of the center of gravity C_5 of the upper five bricks will have $4W$ on the right-hand side, yielding $L/10$ as the maximum offset; then $L/12$; and so on. To get the total offset of the top brick over the bottom brick we add the partial offsets together:

$$L = \frac{L}{2} + \frac{L}{4} + \frac{L}{6} + \frac{L}{8} + \frac{L}{10} + \cdots$$

$$= \frac{L}{2}(1 + \frac{1}{2} + \frac{1}{3} + \frac{1}{4} + \frac{1}{5} + \cdots)$$

The sum in parentheses is the well-known harmonic series. It does not converge; therefore, L—if you add enough bricks—will become greater than any finite number.

16.

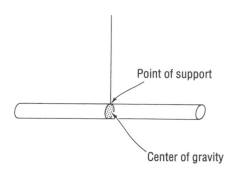

A rod supported at its center of gravity is in equilibrium in any position.

A rod suspended at a point *above* its middle will come to rest horizontally (see diagram). The reason is that in equilibrium the center of gravity must lie on the same vertical as the point of support.

17.

The bolt heads will remain at the same distance. And it doesn't matter which bolt is held stationary.

As long as the threads are meshed, a clockwise movement of bolt *B* around *A*, viewed from the bolt-head end, is the same as a counterclockwise movement of *A* around *B*. While *B* is moving up the threads of *A* toward the head of *A*, bolt *A* is moving down the threads of *B* away from the head of *B*. The movement of the two threads cancels out. (If you don't have two identical bolts at hand, you can get the feel of the problem using one finger of each hand.)

18.

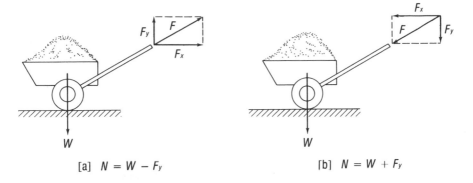

[a] $N = W - F_y$ [b] $N = W + F_y$

Pulling a wheelbarrow is easier. The diagram shows that the pulling force F (a) has an upward component F_y which is subtracted from the weight of the wheelbarrow; this reduces the normal force pressing the wheel to the ground. In (b), pushing produces a downward F_y which adds to the workload. The force of friction $F_f = \mu N$ is smaller when you pull.

19.

The basic equation of motion is

$$s = v_0 t + \tfrac{1}{2} at^2$$

where v_0 is velocity at time $t = 0$. If this initial velocity $= 0$ we have a simplified equation for uniformly accelerated motion from a standstill:

$$s = \tfrac{1}{2} at^2$$

If, again, $a = 0$ we have a simplified equation for motion at constant velocity:

$$s = v_0 t = vt$$

The fallacy in the problem is in taking $\frac{1}{2}at^2 = vt$ because they both "are equal to s." They are equal to s under different conditions—that doesn't make them equal to each other.

20.

Scale C supports only the object, so it reads 30 lb. Scale B supports the object and scale C, so it reads $30 + 2 = 32$ lb. Scale A supports the object and scales B and C, so it reads $30 + 2 + 2 = 34$ lb.

21.

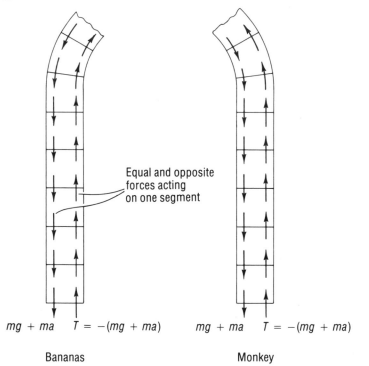

Equal and opposite forces acting on one segment

$mg + ma$ $T = -(mg + ma)$ $mg + ma$ $T = -(mg + ma)$

Bananas   Monkey

The monkey starts pulling himself up by grabbing the rope over his head and pulling down on it. By Newton's third law the rope reacts by pulling the monkey up. The tension in the rope must not only support the monkey's weight but also provide the force for his acceleration up the rope. You can test this statement by holding a weight at the end of a string, then jerking upward. You will feel a sudden added tension in the string as if the weight has become heavier.

Now, the tension is the force exerted by any segment of rope on an adjoining segment. If we divide the rope as in the diagram, we see that by Newton's third law the tension is transmitted along the rope by a series of identical action-reaction pairs. Assuming that the rope is unstretchable, the tension in the rope will be identical all the way along it.

At the monkey's end of the rope the last segment is acted on by the weight of the monkey mg plus a downward pull ma on the rope as he starts the climb. (The pull gives him an acceleration a, so its magnitude is ma.) The downward forces establish a tension $T = -(mg + ma)$ in the rope, which pulls the monkey up.

At the other end of the rope an identical force of tension is acting on the bananas and pulling them up with the same acceleration. The monkey and the bananas will rise together.

22.

Only the forces are identical. In (b) the hand's force accelerates only the cart, but in (a) the force of gravity on the 5 lb weight has to accelerate not only the cart but also the 5 lb weight.

23.

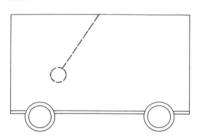

Newton's first law, as stated in the problem, applies only to point particles. If we want to include the motion of bodies with component parts, the word "body" must be replaced by the phrase "center of mass of a system."

Now, first, the components of a system may move while its center of mass remains at rest. This becomes visually more interesting if some components of the system are hidden from view. The figure shows a pendulum cart: a light enclosed cart has a heavy pendulum swinging from its ceiling. When the pendulum is released, the cart rolls back and forth and, if friction is absent, comes to rest where it started. With the pendulum hidden from view we might believe that since the box moves, so does the center of mass of the system. This would disprove the first law as modified above; for, with friction absent, there would be no external force to account for the movement of the cart. But one look inside shows that when the pendulum swings to the right, the box moves to the left, while their center of mass remains at rest.

The second possibility is that the center of mass of a system moves but the box enclosing the inside of the system doesn't. This is the case with our bump-mobile until the hammer hits the plank. When the man starts swinging the hammer, the plank tries to move in the direction of *A* to keep the center of mass at rest, but the external force of static friction acts toward *B* and prevents the plank from moving. This forces the center of mass of the system to move toward *B*. The center of mass is already moving, although the plank and the box are at rest.

When the hammer strikes the plank, it transfers its momentum to the plank. The force of static friction no longer acts; internal motion ceases. By Newton's first law the center of mass continues moving uniformly toward *B;* it would so move but for the force of sliding friction acting toward *A*.

The external force that moves the center of mass of the bumpmobile forward toward *B* is not the strike of the hammer but the static friction which acted when the plank was still at rest. (See illustration on page 12.)

24.

The fluctuations result from the up-and-down movement of the center of gravity of the blood as the heart goes through its cycle. For a person weighing 165 lb the amplitude of the fluctuation is about 1 oz.

25.

More than 3 seconds. The stone keeps losing energy from colliding with particles of air; hence its kinetic energy (and speed) must be less on the downward trip.

26.

1. The stones have the same downward distance to travel. They are acted upon by the same force of gravity. Inescapably, they will reach the ground at the same time.

    2. The force of air resistance F_r is proportional to the square of the speed:

$$F_r \sim v^2 = v_x^2 + v_y^2$$

where v_x and v_y are the horizontal and vertical speeds, respectively. The vertical force on a stone is the resultant of the downward force of gravity and the upward force of air resistance. When a stone is thrown horizontally, its speed is greater and so is the air resistance. Therefore, the stone will fall slower and reach the ground later.

27.

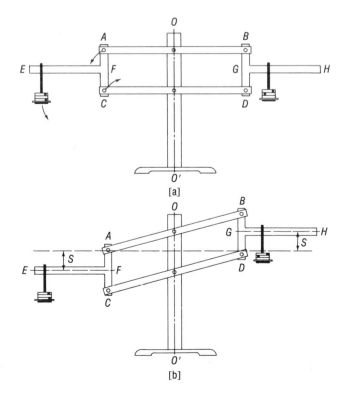

[a]

[b]

In both diagrams, links *AC* and *BD* are always vertical and bars *EF* and *GH*, which are rigidly mounted on the links, are always horizontal. Since *F* and *G* are at the same distance from the central axis *OO'*, the weights on *EF* and *GH* move up and down through the same distance (*S* in diagram b) no matter where they are set on the bars. The weights being equal, the work done by gravity in lowering the weight on *EF* must equal the work that could be extracted from the weight on *GH* after it has been raised. Therefore, neither weight has a mechanical advantage, and the system remains in balance.

If we remove the bars *EF* and *GH* and attach trays to *A* and *B* we obtain a balance with a very useful property: we do not have to be careful to place the object weighed or the weights at the center of the trays. The balance thus constructed is called a Roberval balance after the French mathematician who invented it in 1669.

28.

It would seem that since the left end of the stick is prevented from falling, the right end should descend with a vertical acceleration of less than *g*; thus the object should maintain contact with the right end or even get ahead of it. But this is not what happens.

The reason the falling stick does not violate the law of gravity is that it is not in a state of free fall. If the stick were falling freely with no initial rotation, its center of gravity and every other point would fall with acceleration *g*. But there are two forces acting on the yardstick: an upward force acting on the left end and the downward gravitational force acting on the center of gravity. As a result, there is a force couple trying to rotate the stick clockwise, and there is no reason why this couple cannot produce an acceleration greater than g.

3. Liquids and Gases

29.

Both dams can safely have the same strength. The pressure at any point in a liquid depends only on the height h of the column of liquid above that point, and is equal to $p = p_a + \rho g h$, where p_a is atmospheric pressure, ρ is density of the liquid, and g is acceleration of gravity. The force due to hydrostatic pressure is perpendicular to the surface immersed in the liquid. A single force F applied at a point two-thirds down from the top point of the dam will balance the total force due to the pressure of water.

30.

No. When you drink a liquid through a straw, you don't suck it up: you create a lower pressure in the mouth by expanding the lungs, and the atmospheric pressure pushes the liquid up. In this problem the top of the water is not in contact with the atmospheric pressure, and there is no force available to push the water up the straw.

If the jar is turned over, the water will hardly flow out at all (unless the straw has a large diameter), because once the water starts running inside the straw a vacuum forms above the water in the jar, and the atmospheric pressure tries to push the water back in against the force of gravity. To eliminate the vacuum, you have to let the air in and, simultaneously, the water out. The air will then fill the empty space left by the water. This is why it is better to make two holes in a can if you want its contents to flow out quickly.

31.

Yes; the pan with the bucket on it will go down. The water exerts a buoyant force on your finger equal to $\rho V g$, where ρ is density of water, V is volume of the submerged part of the finger, and g is acceleration due to gravity. By Newton's third law the finger must exert an equal and opposite force on the water. This force is transmitted to the bottom of the bucket and then to the pan of the balance, causing it to dip.

32.

H stands for a hydrogen molecule containing two hydrogen atoms. Thus $H_2 = H$ plus H, not H times H. The same applies to Cl_2 and to HCl.

33.

The ship will float. This is a case of the hydrostatic paradox: The pressure at any point in the water depends only on the vertical distance from that point to the surface of the water. Our natural tendency to think of hydrostatic pressure as having something to do with the total weight of the water in a reservoir is wrong. The pressure at any point on the ship's hull depends only on its depth; the ship "doesn't know the difference" between being surrounded by an ocean and being surrounded by a water layer 1 in. thick. If the dock contains seawater, the waterline will remain where it was at sea.

The horizontal thrust on the hull tends to crush the ship as much in the dock as it did at sea. A ship floating in a small lake has to be as structurally strong as one sailing in the ocean (excluding the effects of ocean waves).

34.

The weight of the tube plus the mercury in it.

The downward force on the top of the tube due to atmospheric pressure is practically equal to the weight of the mercury column in the tube, for the mercury is supported by atmospheric pressure. Now, the upward force on the inside of the tube is zero since, neglecting the pressure of mercury vapors, the pressure above the mercury column is zero. Therefore, the tube is subject to two downward forces: its weight and the unbalanced force due to atmospheric pressure.

35.

Yes. The sponge expands when wet. In addition, the water takes up most of the volume of the wet sponge.

36.

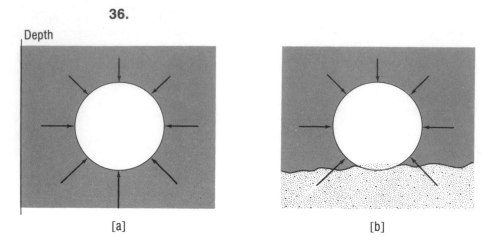

[a] [b]

A submerged submarine is pressed from all sides by forces due to hydrostatic pressure which are perpendicular to the hull at every point and increase with depth (a). Each 10 meters of submergence add 1 atmosphere of pressure.

When a submarine settles on a clay bottom the water layer may be squeezed out from beneath the hull, depriving the submarine of much of its upward buoyant force (b). The downward forces act as before, and in effect they press the submarine to the bottom.

37.

If the pounds don't weigh the same, they must be pound *masses*. A pound (mass) of feathers has much greater volume than a pound (mass) of iron. The buoyant force, which equals volume times density of air, is greater on the feathers; consequently, a pound of feathers weighs less than a pound of iron.

38.

It will bulge toward the larger volume of water. The horizontal thrust on the membrane depends only on the depth of water, and not on its amount in a compartment (Answer 29). If you imagine the larger compartment extended indefinitely to the right, a small volume of water will be able to push back and hold back a whole ocean!

39.

Coffeepots can only be filled to the level of their spouts. In the case given the spouts are equally long, so the pots hold equal amounts of liquid.

40.

No. Check the distribution of pressures in the falling glass of water. For a column of water of depth h from the surface, the equation of motion is

$$pS + ma = mg + p_0S$$

where p is pressure, S is cross-sectional area of the column, and m is mass of the column (given by $m = \rho Sh$, where ρ is density of water). The equation states that the column of water is in equilibrium. The upward forces are pS (the buoyant force on the lower face of the column) and ma (since downward acceleration is observed in the elevator frame to produce an opposite inertial force on the column of water, making it "lighter," as it were). The downward forces are the weight mg and the atmospheric pressure p_0S on the upper face of the column. Solving the equation for pressure:

$$p = p_0 + \rho h(g - a)$$

Now, the buoyant force on the block of wood is $F = \rho V(g - a)$, where V is volume of the submerged part; p_0 cancels out, since it simultaneously produces a downward force on the upper end of the block and adds an equal amount to the upward buoyant force inside the water. The equation of motion for the block is

$$p_0S + \rho V(g - a) + ma = mg + p_0S$$

where on the left we have the upward forces, and on the right, the downward forces.

This gives $V = m/\rho$ for the volume of the submerged part—the same answer we would obtain from Archimedes' principle in an elevator at rest. Consequently, the block will not float higher when the elevator is falling.

We can see intuitively that although the buoyant force is lessened by the downward acceleration, the apparent weight of the block is reduced by the same amount, so the equilibrium position does not change.

(In our discussion we neglect the effects of surface tension. At zero g the block might float clear of the liquid or be sucked under the surface, depending on how much the water wets the wood.)

41.

No. A balloon cannot travel horizontally unless it is blown along by wind. If the wind is steady in one direction, the balloon soon acquires the same speed as the wind. Its velocity with respect to the wind being zero, its occupants do not feel any wind passing them.

42.

The two pails have the same weight. According to Archimedes' principle, a body immersed in a fluid displaces with its immersed part as much fluid (by weight) as the total weight of the body.

43.

The balloon will move to the right, even though the child tends to be pushed to the left. When the car turns right, it is subject to centripetal acceleration acting toward the center of curvature of the turn. Both the air inside the car and the balloon, due to their inertia, want to continue going straight. So to the observer in the car they seem to be thrown to the left, as if pushed by a force, in this case the centrifugal force. Thus there is a pressure buildup toward the outside of the turn as each air parcel presses on its neighbor to the left. The balloon is pushed more strongly to the right than to the left, since the pressure to its left is less than the pressure to its right. Therefore, there is a net buoyant force on the balloon directed to the right, which is exactly analogous to the buoyant force acting in the atmosphere due to the earth's gravitational field.

44.

To the right. The car will not lurch away from the passing cars (as might be supposed), because of Bernoulli's principle: When a fluid and a channel through which it flows are in relative motion, the pressure in the fluid is inversely proportional to the relative velocity of the motion. High velocity makes the pressure less between the cars, leaving the unchanged outside pressure to tend to force the cars together, sometimes creating a dangerous situation.

45.

Most people think humid air is heavier, but it is not. In a humid atmosphere some dry air has been displaced by water vapor, which is lighter than air. Dry air consists mainly (by volume) of 78 percent molecular nitrogen N_2 (molecular weight 28) and 21 percent molecular oxygen O_2 (molecular weight 32). The trace constituents are even heavier. The molecular weight of water vapor H_2O is only 18.

Of course, liquid H_2O is heavier than air and will fall immediately.

46.

Diagram (b) is correct: the middle stream travels farthest to the right. By Torricelli's law, we equate the kinetic energy of a stream at the exit from a hole to the potential energy lost by the water level falling to the level of the hole:

$$\tfrac{1}{2}mv_x^2 = mgh$$

where m is mass of the water, v_x is initial horizontal velocity, g is acceleration due to gravity, and h is initial distance from the top of the water to the hole. Solving for v_x

$$v_x = \sqrt{2gh}$$

Now, the horizontal distance traveled by the stream is

$$s_x = v_x t$$

where t is the time taken by the stream to fall a distance $L - h$ from the hole to the bottom of the can (L being the total initial height of the water in the can). But

$$L \quad h = \tfrac{1}{2}gt^2$$

Then

$$s_x = 2\sqrt{h(L - h)}$$

and this is a maximum when $h = L - h$ (see Answer 1), that is, when $h = \tfrac{1}{2}L$.

47.

An arrow in flight is acted on by three forces: (1) weight, (2) air resistance, and (3) buoyancy of the air.

We can neglect (3), which is very small. (1) acts on the center of gravity, and therefore cannot cause rotation. By elimination, the rotation must be caused by air resistance. The feathered tail causes greater resistance than the sharp point, and thus can align the arrow along the direction of flight, keeping it from turning over and over.

48.

Both phenomena are due to a special kind of friction that exists only in liquids and gases. Friction in the usual sense arises when there is relative velocity, or an attempt at it, between two solids in contact. In fluids, two adjacent layers within a mass of fluid can move at different velocities, giving rise to internal friction—or as it is usually called, viscosity.

Its effect is that the slower layer retards the faster one and the faster one speeds up the slower one. On balance, the kinetic energy of motion is partly converted into heat, and the average motion of the fluid becomes slower.

Perhaps the reader has noticed that a layer of dust on fan blades remains there even after the fan has been churning the air for several hours. This is because the relative velocity of the fluid layer immediately adjacent to a solid surface is zero. The fluid sticks to the surface and cannot slip relative to it. The smallest dust particles will not be disturbed even though the air a fraction of a millimeter away is moving at high speeds relative to the blade.

The most important consequence of the no-slip condition is the formation of a boundary layer, that is, a layer of retarded flow. The process starts with the motionless layer at the boundary exerting a viscous drag on the next layer and gradually slowing it down. As the second layer loses momentum it exerts a viscous drag on the third layer, and so on. Velocity increments from layer to layer become smaller as you proceed away from the wall, because they are proportional to the original velocity differences between layers. We gradually reach a region where the flow is practically unretarded by viscosity, which only acts when there is a velocity difference between two adjacent layers. The first illustration shows what the velocity profile looks like for a flow between two walls. We may think of these walls as the banks of a river to explain why rafts floating closer to the center move faster.

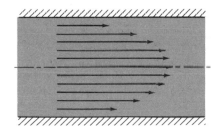

Stationary air also exerts a viscous drag on the water in a river. Therefore, the velocity of water flow is not at a peak at the surface, but at points a short distance beneath the surface. (See the velocity profile in the second illustration.) This means that a heavily loaded raft, being deeper in the water, will be pushed by a faster current and will float faster than a lightly loaded raft.

Air

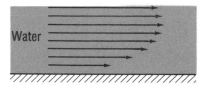

Water

49.

The water resistance is usually less when the stick is held in the moving stream.

Flowing streams are somewhat turbulent. The free stream turbulence induces transition to turbulence in the boundary layer around the stick (see Answer 48). As a result, the slow-moving boundary layer receives extra kinetic energy from the free stream and is able to follow farther around the stick without separation. The form drag is reduced and so is the total drag, since the friction drag is insignificant for blunt bodies.

4. Earth Travel

50.

The frictional force between two bodies is directly proportional to the force pushing them together. When the brakes are applied, the car pitches forward, which adds to the force acting on the front wheels and subtracts from the force acting on the rear wheels. During a panic stop on dry concrete an additional force equal to 10 percent of the weight of the car may be thrown on the front wheels. Since the front wheels support 55 percent of the car's weight to begin with, they now bear 65 percent of the weight, compared to 35 percent for the rear wheels. The frictional force between the front tires and the roadway is twice as great as the force for the rear tires.

51.

Cars are equipped with four-wheel brakes today. We begin with the rear wheels.

If the tires roll without slipping, the friction between them and the roadway is pure rolling friction. The moment you step on the brake the tires are forced to turn more slowly, and for a fraction of a second the translational motion of the car, because of its inertia, is a little faster than the rotational motion of the tires. If the tires are to stay on the car they must slip forward to make up the difference in speeds. At this point the force of sliding friction opposes the slippage by pulling backward on the tires. If the brakes are released the slippage disappears and the car rolls along as before, but at a lower speed.

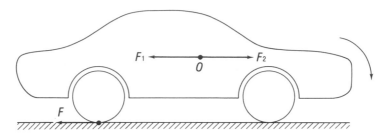

To see what effect sliding friction has on a car during braking, we use a trick which is rather common in mechanics. To O in the illustration (the center of gravity of the car) we apply two oppositely directed forces F_1 and F_2, equal in magnitude and parallel to the force of sliding friction F. The three forces may be considered as a single force F_1 plus a couple, F and F_2. F_1 acts to brake the car, and the couple tries to twist the front of the car downward.

Repetition of this analysis for the case where the brakes are applied to the front wheels again demonstrates a twisting couple acting downward. When the brakes are applied to all four wheels, the couples add up and the twisting effect is reinforced.

52.

The white car will descend with its front forward. The black car will swing around and descend backward.

The basic reason is that the friction between rolling tires and a surface is static friction. (The point at which a rolling tire touches the ground is momentarily at rest, and is therefore in the realm of static friction.) When a tire goes into a skid, as in the toy cars with the locked wheels, static friction gives way to sliding friction. The latter is much smaller—it is easier to keep a car moving once you have set it in motion.

53.

A car descending a hill converts potential energy into kinetic energy. The driver can convert the kinetic energy into heat, and thus slow down his descent, by applying the brakes. Or he can accomplish the same end by coupling the engine to the rear wheels through the gears; friction losses in the spinning engine check the speed of descent. The engine turns fastest when the car is in low gear. The work done against friction per unit time increases with engine speed; hence braking action is greatest in low gear.

54.

Rolling friction occurs when a wheel, ball, or cylinder rolls freely over a surface. Coefficients of rolling friction are generally 100 to 1,000 times as small as coefficients of sliding friction for the same two materials. For a tire rolling on dry concrete the coefficient of rolling friction ranges from 0.01 to 0.03; the coefficient of sliding friction, from 1 to 2.

Before the mid-1950s most scientists believed that rolling friction arises from minute slip between the ball and the surface over which it rolls. But recent studies have shown that such slip plays only a small part in the rolling resistance; this is supported by the fact that lubricants, which greatly ease sliding friction, have hardly any effect on rolling friction.

Let us see what happens when a hard steel ball rolls over a level surface. As it rolls along, it squeezes down on the material around and ahead of it, displacing it upward. If the surface is soft, for example, lead or copper, a permanent groove is produced, as illustrated (a). The force needed to make the groove is almost exactly equal to the observed rolling friction.

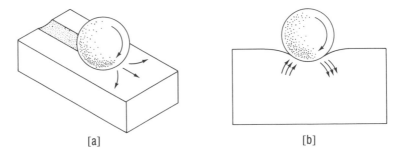

[a]   [b]

If the surface is elastic, such as rubber, a permanent groove is not formed. The rubber behind the rolling ball recovers elastically and pushes it forward (b). No material is ideally elastic, so only a portion of the energy lost in front is recovered in the back. With a very bouncy rubber the losses are small; but if the rubber is rather soggy, most of the energy spent on deformation will be lost, appearing as heat within the rubber.

In rolling, the surface in contact moves mostly up and down like a water bed, and most of the energy can usually be recovered. In sliding, the relative movement is horizontal, and no such recovery can take place, making the losses much greater.

55.

One reason a gas engine needs a transmission is that it develops very little torque (twisting power) at low engine speed. It starts at speeds below 100 rpm but its torque is so small that any appreciable load will stall it. A clutch is needed to disengage the engine from the gears and let it run freely until it reaches 1,000 to 1,500 rpm, at which point it begins to develop useful torque. (A steam engine can develop full power from a standstill.)

56.

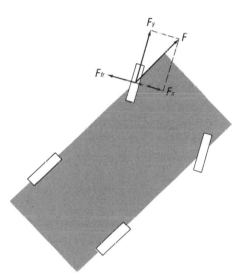

The reader may ask, "Who in his right mind would accelerate while turning?" Of course, the driver slows down enough before turning so that the car will not be thrown off the road when it speeds up around the turn.

In the diagram, when the car accelerates the rear

wheels are in effect giving the front wheels an extra push, represented by the force F. With the steering wheel turned to the left, F can be resolved into two perpendicular components: F_y, which makes the front wheels rotate faster in the direction in which the car is turning; and F_x, which tries to push the front wheels outward. The point at which a rolling wheel meets the road is momentarily at rest, so we are dealing with static friction (Answer 52). The magnitude of static friction increases with F, reaching maximum value just before the car starts to slide. Therefore, as F_x tries to push the wheels outward, the force of static friction F_{fr} acts in the opposite direction, preventing any sideways movement. Thus F_{fr} is the centripetal force that makes a car turn. Frustrated in their attempt to slide in the x-direction, the wheels have no choice but to roll in the y-direction, which is the direction of the turn.

Note that $F_{fr} = F_x$. The more you accelerate, the more you increase the canceling forces, of which F_{fr} is necessary for the turn. Of course, if you reach the point where F_{fr} attains its maximum for the type of road surface, any further acceleration will throw the car into a skid in the x-direction.

57.

The tread on tires slightly reduces their grip on the road under dry conditions. It leaves less rubber in contact with the road. Frictional force is independent of area of contact for rigid solids, but tires are not rigid.

This is also why brake linings are smooth. (They are supposed to remain dry even if the weather is wet.)

When it rains the tire has to penetrate a layer of water or oil and water to reach the solid surface below. A smooth tire would mean the weight of the car would be supported by maximum tire area, making the pressure of the tires against the road smaller, and making it harder for a tire to penetrate the water layer, as it must if the car is not to skid. It is not only that a treaded tire exerts greater pressure on a wet road, but water can then move into the crevices, allowing the raised parts to make contact with the road. On the whole, it is worth sacrificing a little grip on dry roads in order to minimize the danger of skidding on wet roads. And it would not be worthwhile switching tires every time the weather changes.

58.

No. Assuming the cars lose all their speed during collision, the damage is roughly proportional to the total kinetic energy of the two cars which is available for the work of bending, breaking, twisting, piercing, and so on.

Kinetic energy equals $\frac{1}{2}\,mv^2$, where m is mass of the car and v is speed. In case 1:

$$\tfrac{1}{2}m(30)^2 + \tfrac{1}{2}m(30)^2 = \tfrac{1}{2}m \times 1{,}800$$

In case 2:

$$\tfrac{1}{2}m(50)^2 + \tfrac{1}{2}m(10)^2 = \tfrac{1}{2}m \times 2{,}600$$

In case 3 the damage is at the maximum:

$$\tfrac{1}{2}m(60)^2 + \tfrac{1}{2}m(0)^2 = \tfrac{1}{2}m \times 3{,}600$$

59.

While the car turns, a horizontal centrifugal force acts on its center of mass, which is some distance above the ground. Consequently, the car tends to tip to the left; if the turn is fast enough, the inside pair of wheels will leave the roadway.

60.

The moment the wheels lock and start sliding forward—or if they do not lock but slide instead of rolling—static friction between tires and road changes into sliding friction, which is much smaller (Answer 54).

    If there is a sideways force on the car, to the right for example, the sliding friction splits into two components: one acting backward as before, and the other, perpendicular to the first, acting to the left, countering the right-hand drift of the car. Since the resultant sliding friction points in a direction opposite to the instantaneous velocity of the car, the lateral frictional force will be very small. This is because the car's lateral velocity is very slight compared to its forward velocity. Therefore, the road offers hardly any friction in the sideways direction and becomes like ice, but only in that direction. One of the causes of skidding is thus explained.

61.

Steer right at the wall, fully applying the brakes. In the straight stop the kinetic energy of the car will be used up working against friction:

$$\frac{mv^2}{2} = Fx; \text{ or } x = \frac{mv^2}{2F}$$

where F is frictional force and x is stopping distance. If the car brakes from a distance d from the wall, the car will not crash if

$$x = \frac{mv^2}{2F} \leqslant d$$

or

$$F \geqslant \frac{mv^2}{2d}$$

In the circular turn the centripetal force is equal to the frictional force:

$$F = \frac{mv^2}{R}$$

The car will not crash if

$$R = \frac{mv^2}{F} \leqslant d$$

or

$$F \geqslant \frac{mv^2}{d}$$

Since the frictional force required to turn is twice as great, we conclude that to avoid a crash it is better to brake than to turn.

62.

Do *not* attempt to solve the problem as follows: Let x be the speed for the second lap. Then

$$60 = \frac{30 + x}{2}$$

and $x = 90$ mph.

This is wrong. Actually the car would have to drive the second lap at an infinite speed to average 60 mph for both laps.

The average speed $\bar{v}$ is defined by the equation

$$\bar{v} = \frac{2s}{\dfrac{s}{v_1} + \dfrac{s}{v_2}} = \frac{2v_1}{1 + v_1/v_2}$$

where s is length of one lap, v_1 is speed of the first lap, and v_2 is speed of the second lap.

63.

The tangentially mounted spokes of the bicycles have to carry two kinds of load: radial, by supporting the hub which in turn supports the frame and the cyclist; and tangential, by resisting the twisting forces transmitted to the sprocket wheel by the chain (rear wheel) and to the tires by the brakes (either wheel). To be able to carry tangential loads in either direction, the spokes are tangential to the hub in both forward and backward directions.

Wheels with radial spokes did not appear until about 2000 B.C., on chariots in Syria and Egypt. They became universal on carriages and wagons. There the source of locomotion was outside the vehicle, and the wheels carried mainly radial loads.

64.

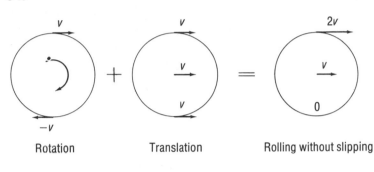

Rotation Translation Rolling without slipping

Relative to the ground, points near the top of a rolling wheel move faster than points near the bottom.

Rolling motion can be diagramed as composed of pure rotation and pure translation. The point touching the ground is at rest. Therefore, it must rotate at the speed $-v$ to give zero total speed when combined with forward translation. At the top the speeds add up instead of canceling, which explains the blurry appearance of the upper spokes in illustrations.

65.

Tires exert much less grip on the road when going downhill or uphill. A tire's grip increases in proportion to the car's weight. When the car is descending only part of the weight is exerted on the road surface, as illustrated. The other part tends to make the car roll downhill. (W is the weight of the car; F_g is the force exerted on the road; and F_r is the force tending to make the car roll downhill.)

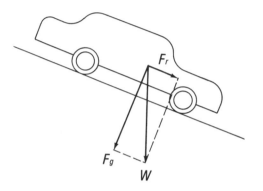

5. Armchair Athletics

66.

Work can only be done if an object is moved in the direction of a force. In the problem both arm and suitcase are stationary. But just because an arm is still doesn't mean its parts can't move and do internal work in the process. Things with internal parts can get tricky: living things are the trickiest of all.

The striated muscles of the human arm are composed of fibers which usually run the length of a muscle. A single nerve cell may control the contraction of up to 1,000 fibers, which form what is called a motor unit. When the nerve cell discharges an impulse, all the muscle fibers in that motor unit contract. The average striated muscle may have about 300 motor units.

The force developed by a muscle depends on the degree of tension in a given motor unit which is in turn determined by how often it is stimulated and, more importantly, on the fraction of the motor units participating in the contraction. At any particular instant various motor units within a muscle are in different stages of their contraction. Some are beginning it, some ending, others are fully contracted, while still others are completely relaxed and resting.

Each motor unit goes through a complete cycle dozens of times a second. As if all that twitching were not enough, at the molecular level there is a furious movement of hundreds of different chemicals carrying energy and materials back and forth from one cell to another.

67.

True. Recent studies report that per pound of lean body weight females are somewhat stronger than males. Greater strength in men is due to differences in weight and fat content rather than to inherent strength of muscle tissue. After puberty, the body weight of females is about 25 percent fat; of males, only 15 percent. Even more important, men weigh more, with the result that women have much less muscle tissue.

Muscle bulk is relatively unimportant, however; athletes use no more than 20 percent of their muscle potential. Today's 14-year-old girls swim faster than Johnny Weissmuller—the original Tarzan—did in the 1924 Olympics, despite the obvious difference in muscle bulk. Women can develop substantial strength through weight training without developing bulging muscles. Apparently, muscle bulk is due primarily to the male hormone testosterone—which is present in women too, but only in very small quantities.

68.

To rise on tiptoe you have to shift your weight forward. With the door in the way, your weight can't be shifted forward.

There is a way to do the trick, though you have

to cheat a little. Take a weight of 10 lbs or more in each hand. After assuming the prescribed position, swing your arms forward. Now you should be able to rise to the occasion!

69.

To keep things simple, let's consider a standing vertical jump, which is only a matter of raising your center of gravity upward. In the first stage you accelerate up from a crouched to a stretched position, raising your center of gravity by a distance s. When the feet leave the ground they cannot do any more pushing off. Accordingly, at this point the center of gravity reaches its maximum speed v_{max}, which can be found from the well-known formula

$$v_f^2 = v_0^2 + 2\,ad \tag{1}$$

Here the initial speed v_0 is zero, and the acceleration a is given by the average net upward force on the jumper, F_n, divided by the mass m of the jumper. The force F_n equals the average force with which the ground pushes back on the jumper (equal and opposite to the force with which he pushes himself off the ground) minus his weight mg. Substituting in (1):

$$v_{max}^2 = \frac{2F_n}{m}s \tag{2}$$

In the second stage the jumper moves upward, powered by his initial momentum, until he momentarily comes to rest in midair, raising his center of gravity from the erect position to the highest point by *h.* Using equation (1) again, we obtain

$$0 = v_{max}^2 - 2gh \tag{3}$$

Combining (2) and (3), we get

$$h = \frac{F_m s}{mg} \tag{4}$$

Now we can understand why small animals can jump so high. Assume that the strength F of an animal is proportional to the cross-sectional area A of its muscles. Then F is proportional to L^2, where L is the animal's linear size. The animal's mass is proportional to its volume L^3. Therefore, acceleration, which equals F/m, is proportional to $L^2/L^3 = 1/L$. Since the stretch distance s is proportional to L, we see from (2) that v_{max}^2 is proportional to $(1/L)L = 1$, that is, the maximum speed is the same no matter what the size of the animal. Therefore, by (3), the height h is also the same. Thus a flea blown up to dimensions a thousand times greater (10 ft length) could still only jump 1 ft off the ground.

Or could it? Such a giant flea would surely collapse under its own weight before it could jump. In blowing it up by a factor of 1,000 we increase its weight by $(1,000)^3$ and its muscle and bone cross section by only $(1,000)^2$. The load per unit area on the poor animal's framework is a thousand times as great. Apparently, science-fiction writers forget this when they try to scare us with giant insects.

70.

When freely bending over you shift your weight forward and simultaneously move the pelvis backward in order to keep the projection of your center of gravity from moving outside your feet, causing loss of equilibrium. With a wall at your back you cannot move the pelvis backward.

71.

The Oslo jumper, who jumped ⅛ in. less, made the better jump. The reason is that the acceleration of gravity g varies across the surface of the earth, depending mainly on latitude and elevation. As you go toward the equator, the linear speed of a point rotating with the earth increases and so does the centrifugal force trying to shake you loose from the earth's surface. As a result, the effective acceleration of gravity g decreases toward the equator. It also decreases as we move up from the center of the earth. Thus g at Oslo (zero elevation, 60° latitude) is 981.50 cm/sec²; at Mexico City (2,240 m elevation, 19° latitude) it is 978.44 cm/sec², 0.3 percent less. This more than makes up for the ⅛ in. difference in the jump (a 0.1 percent difference).

72.

The water pressure on the chest would make breathing through a tube impossible for any length of time, even at a depth of 2 ft. The deepest your head can go and take a single strenuous breath is 3 ft. This demonstrates how overwhelming hydrostatic pressure is, even at moderate depth.

73.

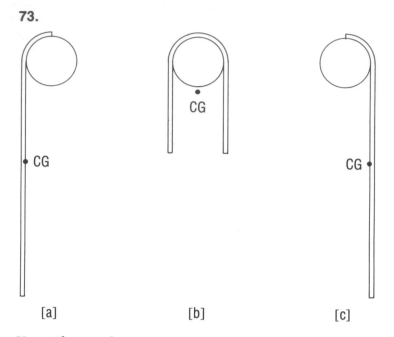

[a]   [b]   [c]

Yes. This is the only way jumpers can clear the bar placed at 7′6″ (229 cm) or higher.

Consider a jumper 6′0″ (183 cm) tall. When he stands erect his center of gravity is 3′7″ (109 cm) above the ground. Now, even the best athlete can raise his center of gravity only about 2′6″ (76 cm). Thus his center of gravity can only be 3′7″ + 2′6″ = 6′1″ above the ground. In executing a 7′6″ jump his center of gravity will pass 1′5″ beneath the bar.

How is this possible? Well, the high jumper has to become like a piece of string. Looking at the successive positions of the string, we see that in (a) its center of gravity is halfway up, in (b) it is at its highest point, and in (c) it has descended to the same height as in (a). The string passed over the bar but its center of gravity passed beneath.

74.

If a child stands on a swing there are many ways he can pump it: what they have in common is that he bends his knees at the end of the backward or forward swing (or even at the end of both), and straightens his knees in the middle of the backward, forward, or both swings, respectively. This has the effect of raising his center of gravity (CG) in the middle of a swing, and lowering it at the end of a swing.

To raise his CG in the middle of a swing the child has to work against two downward forces: (1) the earth's gravitational pull—which increases his potential energy, and (2) the centrifugal force—which increases his kinetic energy. Why the latter? Well, by raising his CG the child gets closer to the point of the swing's support. The angular momentum about the point of support doesn't change at the instant when the CG is raised, since the torque due to a force whose line of action passes through the support axis is zero. The angular momentum equals mvl, where m is the mass of the child, v is his speed, and l is the distance of the CG from the axis. When l is made shorter, v must increase to keep mvl constant. As v becomes larger, so does $\frac{1}{2}mv^2$, the child's kinetic energy. It is the same as when a pirouetting skater pulls in his arms, thus increasing his speed of rotation. When the child lowers his CG at the end of a swing, he loses only potential energy. He can't lose kinetic energy because he is then instantaneously at rest. Thus over a swing cycle there is a net gain in energy, which in turn increases the amplitude of the swing.

75.

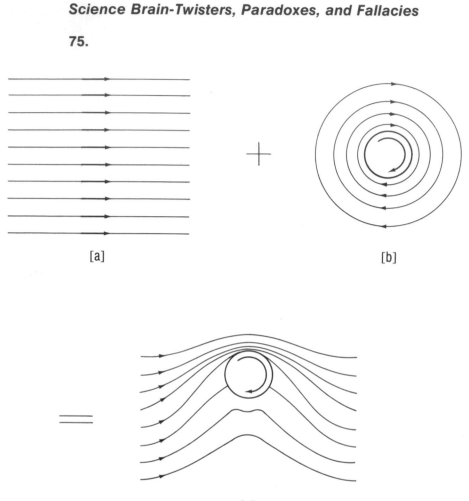

[a]   +   [b]

=   [c]

Lyman J. Briggs's studies in the late 1950s proved that curveballs are no illusion. A good pitcher can make a baseball curve as much as 17.5 in. in the 60 ft between the pitcher's mound and home plate. Such a "perfect curve" travels about 70 mph and spins around a vertical axis at 30 rps. The lateral deflection is generally proportional to the spin and to the square of the linear speed.

Spinning balls, including those used in tennis, table tennis, golf, soccer, and baseball, curve because of the Magnus effect. The flight of a ball spinning, say, clockwise consists of two motions. In (a), if the ball is moving to the left and the air is stationary, an observer flying with the ball will see the ball stationary and the air moving to the right. In (b), the ball is spinning clockwise, making the air in the immediate neighborhood stick to the surface and rotate with it. The speed of rotation of the air in the boundary layer around the ball decreases rapidly with increasing distance from the ball. The thickness of the boundary layer, that is, the layer in which the speed of rotation of the air decreases from the value on the surface of the ball to practically zero, is rather small and depends on how rough the ball is. (If a ball is very smooth, the boundary layer is very thin, and at low linear speeds the lateral deflection of such a ball is usually opposite in direction from that predicted by the Magnus effect. However, even new baseballs are not smooth enough to show this anomalous behavior.)

Combining the two motions by vector addition, we get the pattern shown in (c). At the top the velocities of the streaming and rotating air add to each other; at the bottom they subtract. As a result, by Bernoulli's principle, at the top where the velocity is high the pressure is low; and at the bottom where the velocity is low the pressure is high. The net force due to the pressure difference deflects the ball upward. Similarly, a counterclockwise spin deflects the ball downward.

Why has there been a controversy about curve-balls in baseball while in tennis or golf the lateral curvature is easily seen by anyone? Well, golf and tennis balls are lighter, travel faster, and rotate much faster. A golf ball hit with a 7-iron may attain a spin of 130 rps. But the most important reason is that the time of flight of a pitched baseball is well below 1 second, while tennis and golf balls may be in flight for 2–5 seconds.

76.

The answer is a qualified yes. A curveball travels at a constant linear speed and spins at a practically constant rate. Therefore the Magnus force (Answer 75) due to the spin of the ball produces a small constant acceleration at right angles to the spin axis.

Let us assume a vertical spin axis and a horizontal direction of flight. If the angle between spin axis and direction of flight is less than 90° the force is less, and if the angle is 0° the Magnus force vanishes.

The distance covered in uniformly accelerated motion is $s = \frac{1}{2}at^2$, where a is acceleration. The lateral deflection of a curveball increases with the square of the time. Thus 75 percent of total deflection occurs during the last half of the flight, and 50 percent during the last three-tenths.

The angular velocity at which a curveball deflects from a straight path, as seen by batter or catcher across his line of sight, increases even faster because of the decreasing distance between the ball and home plate. The perspective illusion exaggerates the suddenness of the deflection.

77.

It is erroneous to say *the* reason is that dimpled balls

fly farther. While rough golf balls do, paradoxically, experience less air resistance, the primary purpose of dimpling is to increase the lifting force on a ball given bottom spin.

How does roughness reduce drag? At low speeds it does not; but a full drive sends a golf ball flying at 160 mph. A ball, or any object flying through the air, is enveloped by a thin boundary layer. If the ball is smooth, the boundary layer is laminar, that is, there is no mixing of the sublayers. The main flow separates from the ball, producing a region of back flow and large eddies downstream.

But if the ball is rough, the air in the boundary layer has to go over the hills and valleys. The flow becomes turbulent, which means a lot of mixing and momentum exchange. As a result, the high-speed air flowing outside the boundary layer is able to lend momentum to the low-speed air inside the boundary layer. With this assistance the turbulent boundary layer can flow farther against increasing pressure than the laminar boundary layer can. The main flow remains attached to the ball, making the low-pressure eddying area on the downstream side much smaller than in the laminar case. Moreover, the pressure on the downstream side is not as low. Therefore, the force imbalance between the downstream side and the upstream side of the ball, that is, the form drag, is reduced.

What about friction drag? It increases somewhat with a turbulent boundary layer; however, for an unstreamlined body like a ball, friction drag is insignificant compared to form drag.

The paramount reason for dimples on golf balls is the creation of lift. The ball can impart a spinning motion to only a thin layer of air. Moreover, the (laminar) boundary layer does not follow all the way around the ball. Instead it separates, and it does so earlier on the side spinning against the relative wind, because on the side spinning with the wind the boundary layer is urged onward by the wind flowing past the ball.

A turbulent boundary layer can exchange momentum with the relative wind much more than a laminar boundary layer can. Therefore, the layer on the side spinning with the relative wind will be urged on more than in the laminar case. Its separation point will be moved farther around the ball with respect to the separation point on the side spinning against the relative wind. This increases the force which becomes a lift in the case of bottom spin.

78.

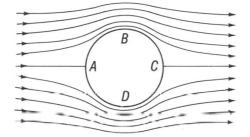

The air drag will be less in the second case by at least a factor of 2. Common sense suggests that a knifelike leading edge will offer less drag since it, so to speak, cuts its way through the air. This is correct—but only when viscous forces are important, as with bodies moving slowly through water. Ships for this reason generally have sharp bows and rounded sterns. But common sense observations cannot be relied on when we enter the realm of high-speed flight.

Suppose we place a circular cylinder in the stream of an ideal nonviscous fluid (first diagram). The particles of such a fluid move along the streamlines without any energy losses due to internal friction, in what is called streamlined or laminar flow. We can see that to the fluid flowing past the cylinder it looks as if the channel is gradually narrowing and then widening again. If the fluid is not piled up in the narrow section it has to accelerate going down the narrowing channel, reach maximum speed at point *B*, and gradually slow down again as the channel widens. Something must be pushing the fluid from behind as it accelerates between *A* and *B*. And something must be pushing the fluid back to slow it down between *B* and *C*. This something is pressure. Without knowing it we have rediscovered Bernoulli's principle, which states that along any particular streamline where the speed is high, the pressure is low, and vice versa. Clearly, our arguments apply equally to the lower part of the cylinder, *ADC*.

Examining the first diagram more closely we see that the streamlines are symmetrical about the line *BD*, so that if the arrows were removed we could not tell whether the fluid was moving from left to right or from right to left. The forward push given to the upstream half of the cylinder, *BAD*, by the accelerating fluid would have to be equal to the backward push applied to the downstream half of the cylinder, *BCD*, by the fluid having to flow against increasing pressure. Therefore, the net force on the cylinder is zero, that is, the cylinder offers no resistance to the moving air. This is known as d'Alembert's paradox.

Our last conclusion cannot possibly apply to real fluids, which stick to the surface of a body, forming a thin coating around it. The next layer rubs against the surface coating, the third layer against the second, and so forth, until the relative velocity of the last layer is practically equal to the velocity of the surrounding fluid. The total frictional force due to all the rubbing is called friction drag, and is one component of the total drag that a body offers to the air moving relative to it. Friction drag is always present, no matter how a body is shaped.

The region in which the relative velocity increases laterally from zero to its free-stream value is called the boundary layer (Answer 48). Since the boundary layer is usually very thin a fluid particle moving inside it experiences the same differences in pressure as those outside it. However, since the fluid in the boundary layer moves much more slowly the particle arrives at point *B* with a lower velocity despite the forward push from the falling pressure.

Starting at *B* the particle must buck not only opposing viscous forces inside the boundary layer but also decelerating forces produced by the rising pressure. The particle gets some help from the slower-moving boundary layer being urged forward by the faster-moving fluid outside it; but when the pressure rises too fast, as when the channel suddenly widens, the boundary layer particles may be stalled.

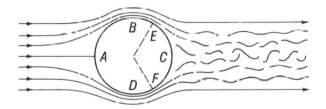

The boundary layer also becomes thicker as the channel sharply widens. The outside flow becomes more distant from the surface of the body and consequently loses some of its ability to urge forward the slowly moving strata deep inside the boundary layer.

Once the boundary layer is stalled, the fluid particles within it may get pushed backward by rising pressure. They collide with the oncoming particles, creating a turbulent region behind the body. The main flow separates from the body and flows around the turbulent region (see second diagram). The pressure there is lower than in the laminar case, so the backward push on the *BCD* portion of the cylinder is less than the forward push on *BAD*. The net force, no longer zero, is called the form drag.

The form drag can be reduced to practically zero by streamlining, where the idea is to prevent separation of the boundary layer by reducing the rate at which the pressure rises in the back. This is accomplished by extending the body gradually to the back so that it tapers off to a knifelike edge. Now we see why the shape of the rear portion of a body is more important than the shape of the front—something common sense would not suggest. (At sonic speeds the shape of the front becomes important too.)

79.

Because 80 minutes = 1 hour 20 minutes.

80.

The airfoil. This streamlined shape is ten times as thick as a round wire, yet produces slightly less drag.

81.

By Newton's third law the rotation of a propeller causes the fuselage to rotate in the opposite direction. A second propeller is used to produce a counteracting thrust in the opposite direction. Similarly, when a conventional aircraft has two propellers on one side, they rotate in opposite directions.

82.

No. More time is lost flying against the wind than is gained flying with the wind.

Suppose that the velocity of the wind is almost equal to the windless velocity of the plane, say 99 mph compared to 100 mph, and that it is 100 mi from A to B. The plane flies at 199 mph with the wind and 1 mph against it. The trip from A to B takes practically ½ hour, but the trip from B to A takes 100 hours!

83.

It seems strange that airplanes take off into the wind, since pushing against the wind slows them down relative to the ground. But what is important in getting off the ground is not the plane's speed relative to the ground but relative to the air. If a plane reaches 100 mph with a 20 mph head wind, its speed relative to the air is 100 + 20 = 120 mph. The plane could go faster, relative to the ground, downwind—say, 110 mph. But then its speed relative to the air would be only 110 − 20 = 90 mph.

84.

False. Actually, an air bump is an upward gust of air that temporarily gives the plane additional lift. A downward gust, producing downward acceleration, is called an air pocket because the passengers feel as if the plane entered a region where there is no air to sustain lift.

Upward and downward gusts generally occur

below and at the cloud level. However, flying may also be bumpy in cloudless air between altitudes of 4 and 7 mi. This is due to clear-air turbulence around jet streams, where rapidly moving air is close to much slower air. Clear-air turbulence is most severe over mountainous areas.

85.

The North American continent is in a zone of prevailing westerly winds, whose average speed near the ground is 5–15 mph. At higher altitudes the westerlies blow much faster, speeds of 50–100 mph being common.

86.

Bumpiness of flight is caused by air gusts, vertical or horizontal, which occur especially during squalls and thunderstorms. Smaller gusts may result from convection currents, that is, rising currents of warm air and descending currents of cool air.

The reason for convection currents is that certain kinds of surfaces are more effective than others in heating the air directly above them. Plowed ground, sand, rocks, pavement, city buildings, and barren land give off a great deal of heat, whereas water and vegetation tend to absorb and retain heat.

As the rising air cools its density increases until it reaches a point where it becomes denser than the surrounding air and begins to descend. Thus rising currents of air cannot go higher than a certain altitude (which varies depending on local conditions). Above this altitude the flying is smooth.

7. Sounds and Voices

87.

Yes. The grasshopper, by rubbing its legs against its rough abdomen, can produce sounds ranging from 7,000 to 100,000 vibrations per second (Hz, after Heinrich Hertz). But it can only hear sounds ranging from 100 to 15,000 Hz.

Most animals can hear much more of the frequency range than they can produce. From the point of view of communication it would be silly for them to make sounds they cannot hear. On the other hand, it is important for animals to hear the sounds produced by their predators. As a result, they can hear many sounds not in the range they themselves produce. A few examples are given in the table (frequencies are in Hz; C_4 is middle C).

	Emission Range	Reception Range
Dog	452– 1,080	15– 50,000
Cat	760– 1,520	60– 65,000
Robin	2,000– 13,000	250– 21,000
Porpoise	7,000–120,000	150–150,000
Bat	10,000–120,000	1,000–120,000
Piano	28– 4,186	—
Pipe organ	10– 8,000	—

	Emission Range			Reception Range
Telephone	250–	2,800		—
Hi-fi system	15–	30,000		—
Man	80–	1,200		16– 24,000
Bass	82.4–	370	$(E_2 - F\#_4)$	
Baritone	110–	392	$(A_2 - G_4)$	
Tenor	123.4–	440	$(B_2 - A_4)$	
Alto	174.6–	587	$(F_3 - D_5)$	
Mezzosoprano	220–	698	$(A_3 - F_5)$	
Soprano	261.6–	880	$(C_4 - A_5)$	
Coloratura soprano	261.6–	1,175	$(C_4 - D_6)$	

88.

The difference is real, and would still be there even if the tape recorder did not introduce any distortions. Others (including people and tape recorders) hear us differently from the way we hear ourselves.

When we hear ourselves speak, the sounds reach us in two ways: by air and through the skull. When we click our teeth or chew a cracker the sounds are transmitted mainly by bone conduction. The same happens when we hum with closed lips. (If you stop your ears with your fingers, the hum will sound louder.) In the air-conducted sounds of our speech some of the low-frequency components are, unfortunately, lost. Most of the vibrational energy of speech goes into components with frequencies above, say, 300 Hz, relatively little being low-frequency. This is why our voice sounds thinner and less powerful to others than to us, since we have the benefit of hearing it through bone conduction as well as air conduction.

89.

The prongs of a vibrating tuning fork move in opposite directions: both outward or both inward. The
center of mass of the tuning fork remains at rest.

If the tuning fork had only one prong, its center
of mass would move with the vibrations of the prong.
A massive base or strong force would be needed to
minimize the energy transfer to the support. If you
tried to hold a one-prong tuning fork in your hands or
mounted it on a small base, it would lose its energy
to your hands or the base, and its vibrations would
die out quickly. An analogy can be drawn using two
balls. Ball A strikes ball B, which was initially at rest.
If A is much lighter than B, it will lose very little of
its energy and simply be deflected back with almost
the same speed. However, if A and B have identical
masses, A will transfer all of its energy to B and come
to a stop while B will take off with A's original speed.

Thus, two prongs of a tuning fork make a heavy
base unnecessary. The tuning fork can be held by
hand, and transfer its energy only to the air by producing a sound wave. Since the area of the tuning
fork is small, the fork loses its energy to the air
slowly, and therefore can produce a steady tone for
many seconds.

90.

The grasshopper does not move all the surrounding
air at once: it squeezes the air close to it during each
vibration. The squeeze does not stay frozen but
spreads in every direction away from the grasshopper, because air is elastic. When squeezed it bounces
back, in the process squeezing the air around it. The

latter also bounces back, in turn squeezing air even farther away, and so the squeeze or the compression travels outward. When the compression enters our ears, we perceive it as sound.

The amount of squeeze is very small. Pressure variations due to audible sound waves are between 0.0002 and 1,000 dynes/cm². For comparison, atmospheric pressure is about 1 million dynes/cm². In ordinary conversation the pressure 1 meter from the speaker's mouth changes by only 1 dyne/cm².

91.

A sound wave (or any wave) is partly reflected and partly transmitted whenever it encounters a change in the resistance to its motion. A sound wave is reflected from a solid wall because of a sudden increase in density. The compression part of the wave is reflected as a compression, and the rarefaction part as a rarefaction. When a compression traveling down a narrow tube comes to an open end, it expands outward, thus creating a deficit of pressure—that is, a rarefaction. Air to the left of the rarefaction is sucked in to fill it. Thus the rarefaction travels to the left, even though the air parcels themselves travel to the right, taking their randomly moving molecules with them. Similarly, the rarefaction part of the wave is reflected from the open end as a compression.

92.

1. Sound travels faster in warm than in cold air. This is because a sound wave is transmitted by molecules

bumping into their neighbors: in warm air the molecules travel faster so they can get to their neighbors sooner, enabling the compression to move at a higher speed. The temperature dependence of the speed of sound in dry air is given by

$$v_t = v_0 \sqrt{1 + \frac{t}{273}}$$

where v_0 (331.3 m/sec or 1,086.7 ft/sec) is speed of sound at 0^0 C and t is Celsius temperature. At room temperature of 20^0 C (293 K) the speed of sound is 344 m/sec.

Now, on a calm, clear night we often have a temperature inversion, in which temperature increases with height, up to a certain elevation, because at night the ground radiates its heat to the air and no longer receives heat from the sun. The heat is radiated faster if there are no clouds to stop it. The air warmed by the ground goes to higher elevations, and stays there if there is no wind to mix it back with the colder air beneath.

Suppose the temperature increases rapidly from 10° C at the ground to a maximum of 15° C at 20 m elevation. If you cup your hands around your mouth and shout to a friend some distance away, your mouth produces a sound beam with a nearly flat wave front. As the beam travels at an angle to the ground, the top of the wave front is moving through warmer air than the bottom and thus travels faster and overtakes the bottom. This has the effect of tilting the wave front downwards.

It is like paddling a canoe—if you pull harder on the right-hand oar the canoe will turn to the left. The effect can be demonstrated at home by letting a spool roll at an angle from a smooth surface onto a rough one. At the boundary the spool will change its direction away from the smooth surface.

Similarly, sound waves curve away from warmer air. The sound beam will in effect be reflected from the top of the inversion layer and directed toward the ground, where it may again be reflected and sent upward at the same angle as before. The sound is limited to travel within a thin sheet of air—essentially a two-dimensional space, so that the intensity of sound decreases in proportion to only the first power of distance rather than the second power of distance as is the case in ordinary three-dimensional space. If the sound is confined to a beam, its intensity will diminish even more slowly. (However, due to their longer wavelengths sound beams spread more readily than light beams.)

2. In summer the water is usually cooler than the air, causing a temperature inversion, with lower air temperature near the water. A sound wave emitted near the water will curve back toward it. The surface of water, being very smooth, is a better reflector of sound than soil. The sound wave will be mostly reflected, then bent again, reflected, and so on.

3. In normal atmospheric conditions temperature decreases with altitude. As a result, sound waves originating near the ground bend away from it and toward a balloon floating overhead, while sound waves produced by the balloonist also bend away from the ground, often missing it completely.

Also, the balloonist produces sounds in air of slightly lower density than on the ground; as a result, the energy of his sound waves is less than the energy of the sound waves produced by people on the ground. Air always finds it easier to move from regions of higher density (higher pressure) to regions of lower density (lower pressure). (This effect is rather small: the difference in air densities is 10 percent at most.)

Finally, the balloonist is in a region of silence where faint sounds are easily heard, while people on the ground are immersed in a flood of sound, making the balloonist's voice hard to pick out from the background noise. In fact, the balloonist can often hear an echo of his own voice from the earth, while people on the ground pay no attention to his calls. (The echo is loudest over still water and weak over freshly fallen snow.)

4. This used to be explained by noting that upper winds may blow in a direction opposite to lower winds. If there is a westerly wind below and an easterly wind above, points west of the sound source will be in a zone of silence (Answer 93), since the sound will be deflected upward. Then, as the sound reaches the upper wind which blows from the east, the sound will be deflected back to the ground.

This is correct in some cases, but by no means explains how, often, the zone of silence surrounds the source completely at some radius from it, and the sound is heard in several directions beyond the zone.

The accepted explanation today is that the return sound is due mainly to a temperature inversion high in the atmosphere. The firing of a cannon sends out a hemispherical wave which expands as it rises above the ground. If the air temperature decreases with height, as it usually does, the wave bends away from the ground. Enough sound is usually diffracted back to the surface, especially at lower frequencies, so that the cannon fire can be easily heard over a considerable area around the source. But as the wave travels upward, the diffracted sound finds it harder and harder to reach the ground because of increasing distance, and beyond a certain radius around the source there is a zone of silence.

When the sound wave reaches a height of 10–15 km the air temperature stops decreasing and begins to increase slowly with height to a maximum at about 50 km; the reason is absorption of the intense ultraviolet radiation from the sun by the ozone layer—which protects us from having our skin burned to a crisp by the sun. (Some ultraviolet gets through; otherwise we could not get a tan.)

On meeting warmer air, the sound wave bends away from it and travels back toward the ground. Only a few waves manage to survive such a long journey. The intensity of the sound continually decreases due to spreading into space and absorption by the air. It is only the sounds of large explosions and artillery fire that manage to make it back. A lot also depends on favorable atmospheric conditions.

93.

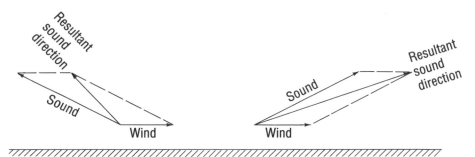

The wind cannot blow the sound back, unless it reaches the speed of sound. (If the wind blows at 50 mph, the sound will be carried downwind at 770 + 50 = 820 mph, where 770 mph is the speed of sound. Upwind, net speed is $770 - 50 = 720$ mph.)

The explanation is that the wind lifts the sound up on the upwind side so that it goes over your head. In the first diagram the velocities of the sound and wind add vectorially. (The length of the wind velocity vector is exaggerated.)

On most days air temperature decreases with elevation. (The air is mostly heated not directly by the sun but by the ground.) Since sound waves curve away from warmer air (Answer 92), the pattern of sound rays emitted by a point source located off the ground appears as in the second diagram, assuming there is no wind. The black areas on either side rep-

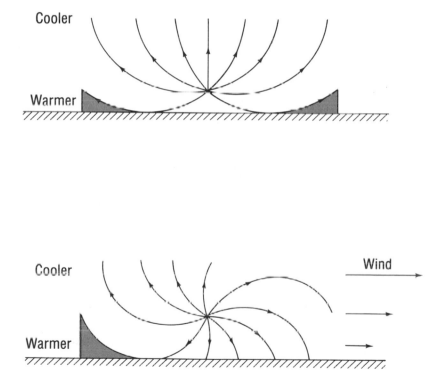

resent sound shadows, that is, regions where very little or no sound is heard.

Also, wind speed greatly increases with elevation. To show the effect of wind we have to combine the sound velocity and wind velocity vectors at each point, taking the increase in wind speed into account. The result is shown in the third diagram. There is a definite sound shadow on the upwind side. The shielding is not complete, because sound diffracts into the shadow, especially at lower frequencies; but the high-frequency sounds, including the high-frequency components in speech, are effectively screened out. Since it is the latter that make speech intelligible, the wind causes problems not only in hearing speech but also in understanding it.

94.

To keep the sound produced on the back side of a loudspeaker cone from canceling the sound produced at the front.

Every time the front of the cone moves forward, producing a compression of the air, the back side moves backward, producing a rarefaction. Thus the sound waves produced by the two sides of the cone are opposite in phase. The back sound, spreading in every direction, travels around the cone to the front and meets the front sound. At low frequencies, where the sound wavelengths are as long as 10 ft, the distance traveled by the back sound is small compared to the wavelength. It arrives at the front with almost the same phase that it started with. Now a compression meets a rarefaction, and the bass all but disappears.

A solid board (a *baffle*) increases the distance

the back sound must travel to reach the front. As a result, the frequency at which destructive interference occurs is reduced, bringing back some of the bass. If the speaker sits on the floor, then at least the back sound cannot get around one of its four sides, reducing the amount of cancellation by one-fourth.

95.

Freshly fallen snow is like the sound-absorbing material often seen on office ceilings—it has billions of tiny holes between snowflakes into which the air molecules carrying sound energy penetrate and get lost in a gigantic maze of microscopic tunnels. The sound wave cannot find its way out and vanishes, its energy being converted into heat.

The blanket of silence that comes with snowfall can be dangerous, concealing sounds of human habitation and making it hard to locate people buried under snow.

96.

In the temperate zones, in both northern and southern hemispheres, the prevailing winds are from the west. If the freeway is built west of the city, the winds will carry automobile exhaust and noise right into people's windows. (The same considerations apply to siting industrial plants.)

Travelers can observe that the more expensive sections of big cities, like Georgetown (Washington, D.C.) and Beverly Hills (Los Angeles), are located west of the downtown areas. In 19th-century Europe factories and smoke-belching stacks were built right in the middle of cities. Air quality and noise levels in the eastern sections were often intolerable.

97.

The prongs produce sound waves of opposite phase.

Suppose the tuning fork vibrates in a plane perpendicular to the plane of the ear. As one prong moves toward the ear, producing a compression, the other moves away from the ear, producing a rarefaction. The prongs being a small fraction of a wavelength apart, as noted, the waves will meet with still almost opposite phases; they will practically cancel each other, resulting in a very faint sound.

Similarly, if the fork vibrates in a plane parallel to the plane of the ear, the fork behaves as a single source of sound. Rarefactions and compressions reinforce each other to produce a relatively loud sound.

The rotating fork goes through the two positions described plus all the positions between. The sound will vary from loud to faint, going smoothly through the intermediate intensities.

98.

The main reason is that lightning follows a sinuous path. Some points on its path will lie closer to the observer than others, so the sound of thunder will be extended. If the nearest point is 5,000 ft closer than the farthest point, the thunder will roll about 5 seconds, since the speed of sound in air is about 1,000 ft/sec.

Also lightning often consists of many strokes following each other in rapid succession. Thirty to forty strokes have been observed along much the same path at 0.05 second intervals. The sound waves produced by multiple lightnings interfere with one another, resulting in thunder that intensifies and diminishes.

Then too, the sound wave produced by lightning is bent by and reflected from layers of air at different densities. Warm air (usually found near the ground in the daytime) has a lower density than cold air (Answer 45). Or layers of warm air may alternate with layers of cool air, with their relation constantly changing, since air is always in motion. (The sound of thunder may be carried 25 mi, although 10 mi is the typical distance.) A similar effect has been observed in the canyons of the Colorado River, where the sound of an outboard motor, echoing off the canyon walls, can sometimes be heard 15 minutes before the raft it is on arrives and 15 minutes after it passes.

8. Heat

99.

Pour half of the cold water into container *D* and place *D* in *A* (which contains the hot water). The final temperature in both *A* and *D* will be 60° C. Now pour the water at 60° C from *D* into *C*, and repeat the first procedure with the other half of the cold water and containers *D* and *A*. The final temperature of the water in *D* (containing the rest of the originally cold water) placed in *A*, and of the water in *A* will be about 47° C. Pour the water in *D* into *C*; the final temperature of the mixture will be about 53° C. The cold water has been warmed to 53° C by cooling the hot water to 47° C.

100.

Who needs boiling water at a temperature of, say, 70° C? You could safely put your finger in it. Meat, eggs, and vegetables would take forever to cook at such a low temperature. Medical instruments could not be sterilized, and so on.

101.

First, 37° C is the temperature inside the body. The rate of heat loss by radiation from the skin is determined by the temperature difference between the air and the skin, and skin temperatures are usually much lower (back, 32° C; legs, 29° C; feet, 10° C or lower).

Second, still air is a poor conductor of heat. If you touch a metal object in a room it will feel cold even though it has the same temperature as the air, because metals conduct heat better than air does.

Third, heat is lost by evaporation of water from the surface of the skin. If the air is still, a stagnant warm-air layer forms over the skin. The layer is saturated with water vapor, making evaporation slow. If wind or a fan breaks up the stagnant air layer evaporation proceeds quickly, making us feel pleasantly cool or ice cold as the case may be. Even in a gentle 2–3 mph breeze the skin cools down twice as fast as it does in air moving less than 1 mph. This is called the wind-chill effect.

102.

Most of us think of ice as inherently slippery. Our ignorance is understandable, since nobody in his right mind goes ice-skating when the temperature drops to −10° or −20° F.

It is only possible to skate on ice because the ice melts temporarily under the sharp runners, creating a thin lubricating film of water between skates and ice. Pressure and friction are responsible. Pressure lowers the temperature at which water freezes, making it possible for ice to exist as water below 32° F. Friction produces heat that helps melt the ice.

But when the temperature drops below the point at which ice can be pressured into being water, there is no lubricating effect, and skating becomes difficult indeed.

103.

Wait 5 minutes before adding the cream. According to Newton's law of cooling, the rate of cooling is proportional to the difference between the temperature of the cooling body and that of the surrounding medium. The temperature T at a time t *is*

$$T = T_a + (T_0 - T_a)e^{-At}$$

where T_a is air temperature, T_0 is temperature of the body at $t = 0$, and A is a constant depending on the size, shape, and composition of the body. The formula is only approximate, but is useful in practical calculations.

If you begin by cooling off the coffee with cream, the difference between its temperature and that of the air is decreased, and it will cool off more slowly.

104.

A thick glass is more likely to break. Glass is a poor conductor of heat. When you pour hot water into a glass, its inner layer is exposed to heat and expands immediately, but the outer layers begin to expand only after the heat has passed through the thickness of the glass. The inner layers expand while the outer ones are still the same size, which sets up enormous stresses across the glass that disappear only when the whole glass is expanding uniformly. Heat takes longer to pass through a thick glass, of course.

105.

The freezing compartment cools the air around it. Cool air is denser than warm air; by Archimedes' principle, it loses its buoyancy relative to warm air and gravitates downward. The warm air is displaced upward, is cooled by the freezing compartment, and gravitates downward in turn. Thus convection currents are set up inside the refrigerator which carry cold air to the bottom sections.

If the freezing compartment is placed at the bottom, the air cooled by it will stay there without helping to cool the food upstairs.

106.

Certainly. Ice is a solid and its temperature can be lowered like that of any other solid.

When the temperature outdoors is $-20°$ C, the top layer of ice on a lake is at $-20°$ C. The bottom layer of ice is in contact with water. Now, water becomes lighter as its temperature decreases from 4° C to 0° C; therefore, the water temperature in a lake ranges from 4° C near the bottom to 0° C just below the ice. The bottom layer of ice will be near 0° C. This large temperature gradient across the ice crust is possible because ice is a poor conductor of heat. (If it conducted heat well, the Eskimos would be shivering in their igloos!)

107.

When the surface of the skin is exposed to cold weather, the blood vessels in the skin contract, cutting back blood flow to the skin and sending more blood to the inner parts of the body, thus reducing heat loss. This occurs primarily at the hands and feet; as a protective measure the body never cuts back the flow of blood to the head. As a result, the body's biggest heat leak is through the top and back of the unprotected head. The Eskimos know what they are doing when they wear thick furry hats.

108.

Yes. A vapor is a gas below its critical temperature. For water the critical temperature is 374° C. Above this temperature steam will not condense into water, no matter how much pressure is applied; such steam can exist only as a gas. (Steam is defined as water vapor at or above the boiling point of water, that is, 100° C at normal pressure.)

109.

On today's cars radiators have a tight cap that lets vapor pressure build up. The greater the pressure, the higher the boiling point: the radiator's water will not boil at 100° C (212° F) but perhaps at 120° C. Since the temperature of the water can safely go up to, say, 100° C, the radiator can be smaller because the water in it does not need to be cooled as much.

    When the pressurized cap is removed, the pressure in the system suddenly falls to atmospheric value. If the water is already at 100° C it will instantly turn into steam, expanding in volume 1,700 times. An eruption follows.

110.

Yes. Most of the water in the tea eventually oozes through the pores in your skin and evaporates. Each gram of evaporating water takes 539 calories of heat away from the skin. If the temperature of hot tea is 85° C, the heat received by your body as the tea cools in the alimentary canal to body temperature of 37° C is only about one-tenth of the heat lost through evaporation.

111.

For the same reason that it is almost impossible to make snowballs on a very cold day. The pressure under your shoes lowers the melting point of ice, which means that water can exist at temperatures below 0° C (Answer 102). As a result snow crystals become covered with a thin lubricating layer of water, which eliminates the characteristic squeaking caused by high friction between snow crystals as they slide past each other.

112.

The amount of heat produced by an animal is roughly proportional to its volume L^3 (where L is the animal's linear dimension), but the amount of heat lost by an animal is proportional to its surface area L^2. The ratio of heat production to loss is $L^3/L^2 = L$. The smaller the animal, the greater the heat loss.

This explains why many arctic species are larger than their near relatives in warm climates, and why flies grow larger in winter.

Small animals are also disadvantaged because they are less able to carry thick fur. Unfurry animals such as weasels or mice spend the winter in sheltered runways under the snow, only rarely coming to the surface.

113.

Certain solid substances have a strong tendency to attract gas molecules to their surface (which is called adsorption). When a piece of charcoal is activated (heated to remove adsorbed matter) it is able to remove large quantities of poisons or impurities from a stream of air, and for this reason is used in gas masks and cigarette filters.

Dry coffee grains are similar to charcoal in having a porous surface, and therefore a large surface area. The adsorptive capacity of substances decreases with increasing temperature, so when coffee grains are placed in hot water, they release a large fraction of the gases adsorbed to their surface. The gas is released in the form of bubbles, and this looks very much like boiling.

114.

When salt or any other substance is dissolved in water, the freezing point is lowered. This is why in winter salt is sprinkled on highways and sidewalks. A salt and water mixture does not freeze above $-21°$ C.

The molecular mechanism of the phenomenon is interesting. Freezing and thawing occur simultaneously in different parts of the system. The solubility of salt in ice is very low; also, a piece of ice is always covered with a very thin layer of water, in a room at normal temperature. So, when we sprinkle ice with salt, the salt remains almost completely in the water layer, forming a highly concentrated solution.

At the boundary between ice and water some H_2O molecules are leaving ice and entering water (melting) and others are leaving water and entering ice (freezing). When salt dissolves in water it dissociates into sodium and chlorine ions, Na^+ and Cl^-, which are electrically attracted to the hydroxyl and hydrogen ions, OH^- and H^+, respectively, in the water. The ions combine; as a result the number of water molecules in the salt solution is reduced. Then fewer water molecules are available to enter the ice, and the freezing process slows down. The melting process, on the other hand, is little affected by salt. To speed up freezing so it balances melting again, the temperature must fall.

In the parlor trick, a close look shows that the salt melted some ice all around the edge of the match. Melting takes energy, and the energy was obtained from the film of water under the match (which was protected from the salt at the time of sprinkling). The removal of heat from the water causes it to freeze, thus cementing the match to the ice cube.

115.

A piece of ice is normally covered with a thin layer of water, even at temperatures below freezing. When the water molecules in that layer find themselves between two ice surfaces, more of them leave the water and attach themselves to ice (that is, freeze) than leave the ice and wander off into the water (that is, melt). Bonds form between the ice cubes, giving rise to what is called ice sintering.

116.

True. A gallon is a unit of volume. The U.S. gallon equals 3.7853 liters (231 cu. in.). Gasoline, like most liquids and solids, expands on heating, and it does so at the rate of 0.6 percent for each 10° F of increase. So, when you buy a tank of gas on a hot summer day, each gallon contains a bit less gas by weight than on a cold day.

A standard gallon of gasoline is defined as 231 cu. in. at 60° F. Next time you buy gas, take a thermometer. If the temperature of the gas is above 60° F, ask the dealer to give you a refund!

117.

An enclosed space above the surface of a liquid contains its saturated vapor. When the liquid reaches boiling point, the vapor pressure becomes equal to atmospheric pressure. As the piston is slowly raised, the space above the boiling water becomes quickly filled with saturated water vapor at atmospheric pressure. This balances the pressure exerted on the water outside of the tube, and the water level in the tube remains unchanged.

If you raise the piston quickly, water vapor will not form fast enough to fill all the space between boiling water and piston. The vapor will violently expand, cooling down in the process. Because of the rapid production of vapor, the surface water layer in the tube will also become cooler. Therefore, the vapor pressure in the tube will be less than atmospheric, and the water will rise to a height such that the sum of hydrostatic water pressure and saturated vapor pressure equals atmospheric pressure.

As the heat is supplied to the water the surface layer will resume boiling, raising the temperature of the saturated vapor back to 100° C and its pressure back to atmospheric. Then the water level in the tube will descend until it is level with the rest of the beaker.

118.

No. The type metals of the printing industry, which are lead-tin-antimony alloys, also expand upon solidifying: as a result, they fill every nook and cranny of the mold. The same is true of sterling silver alloys, which are often cast into intricate designs, especially for the handles of tableware. Silicon and germanium also expand when they freeze.

Water is the only known substance that expands during the last few degrees above freezing point.

119.

The temperature of air from your lungs is 37° C, while the temperature of the skin on your hands is only 25°–30° C. On blowing gently you bring the warm air from the lungs in contact with the cooler skin, producing a feeling of warmth.

When you move your hand farther away, the room air at 18°–22° C has more chance to get sucked into the stream of air from the lungs. This occurs by Bernoulli's principle (Answer 44), for the air in a stream is always at lower pressure. If you blow hard the pressure is lowered further, sucking more room air into the stream so that the air that reaches the hand is well mixed with cool room air, producing a feeling of cold.

It is also important that a stream of air breaks up the stagnant air layer above the surface of the skin, which is saturated with water vapor emanated by the skin. Evaporation speeds up, taking away large quantities of heat, and resulting in a sensation of cold. The harder you blow the greater the effect is.

9. Electricity and Magnetism

120.

There is no voltage on you. A person standing out-doors is a grounded conductor, and his skin is an equipotential surface like the surface of any conductor. The voltage on his skin is everywhere the same and equal to the voltage of whatever he is touching. But he is touching the ground, which has zero voltage, so he is in no danger.

121.

Injury from electric shock depends mostly on the amount of current that flows through the body, not on the voltage applied. Current is higher the lower the resistance put in its path—this is why shoes with thick rubber soles make good sense when working with electrical equipment. Thick rubber puts an al-most infinite resistance in the way of the current, which, unable to make a complete circuit—in through the hand and out through the feet—will not flow at all.

The resistance of the human body has two parts: contact resistance, and resistance of the body proper. Contact resistance, which resists the attempt of the current to enter the body, depends largely on how wet the surface of the skin is and how thick the top layer of the skin is. A dry callused hand may offer contact resistance of 10^5 ohms/cm²; if wet, its resistance may drop to 1,200–1,500 ohms/cm². This is why most electrical injuries occur in bathrooms.

The resistance of the body itself is only about 200 ohms, since the body is mostly made up of water.

Current density has some effect too. Small animals are more susceptible to electric shock than large ones, since in the latter the current is distributed over a greater volume. This may explain the belief that thin people are susceptible to electric shock.

At below 1,000 volts alternating current is more deadly than direct current; at higher voltages, the reverse is true. An alternating current is especially capable of upsetting the control center of the heart, which either stops or goes into uncoordinated contractions, called fibrillations, and fails to pump blood. A low voltage may also produce a steady muscular contraction of the hand muscles so that you cannot release your grip. At high voltages, the muscular contractions are so violent that you may be thrown clear.

On the other hand, a 110 volt line will not reach for you with a spark, but a 50,000 volt power line will spark over to you before you touch it.

122.

The resistance of a wire decreases as its temperature decreases. Electrons find it easier to flow through a crystal lattice whose atoms vibrate less energetically.

As a result, the resistance of the cooled part will decrease while that of the other part will not. The total resistance is smaller, and by Ohm's law ($I = V/R$) the current flowing through the wire is larger, since V is assumed constant. Now, the heat produced by a current I is I^2R per unit time. The resistance R of the uncooled part is the same, but the current flowing through it is larger, so the heat released will be much larger than before cooling. Even a small difference in I will have a measurable effect, since the amount of heat is proportional to the square of the current.

123.

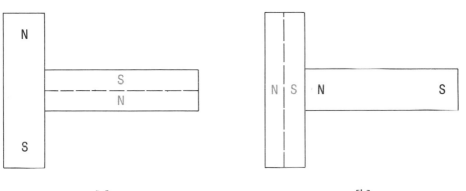

[a]   [b]

Place the bars as shown. There are two possibilities; in each, the magnetized bar is shown with an N and S at its ends. In (a) there will be no net attraction between the bars. The forces between the poles of the magnet and the induced poles of the bar are equal and opposite, canceling each other out.

In (b) there is a net attraction because the attractive force between N in the magnet and the induced S in the bar is much larger than the repulsive force between N in the magnet and the induced N in the bar. The forces between S in the magnet and the induced N and S in the bar pretty much cancel out, since the relative difference in distances between the poles is small.

124.

The orientation of a transmitting or receiving antenna depends on the polarization of the electromagnetic wave used. If the wave is horizontally polarized, the electric field vector vibrates in a horizontal plane. The magnetic field vector, always perpendicular to the electric field vector, vibrates in a vertical plane. The orientation of the electric field vector is the same as the direction of the AC in the transmitting antenna wire. If the wire is horizontal, the wave emitted is horizontally polarized.

In the United States and most other countries, horizontal polarization of television and FM waves is considered superior. In England vertical polarization was thought best.

125.

No. The electrons in the wire are at rest relative to the walking observer, but the positive ions in the crystal lattice move in the other direction at the same speed. Since a conductor as a whole is neutral, the number of positive ions and electrons is the same. Moreover, a positive charge moving in one direction sets up the same magnetic field as an equal negative charge moving in the opposite direction at the same speed; so the positive ions moving backward set up the same magnetic field as the electrons moving forward.

If you walk at, say, half the electron speed, the

magnetic field due to electrons moving forward will be half as strong, but the contribution from the field of the positive ions moving backward at half speed will give the same magnetic field as that experienced by an observer at rest. You can try any speed and reach the same conclusion. There is no way to move to make the magnetic field vanish.

126.

The image on a TV screen is produced by a scanning electron beam. At any moment the beam strikes only one point on the screen. (Color TV sets have three beams emitted by separate electron guns that correspond to the three primary colors; but even in color sets the beams strike three color dots within a triangular area so small that it can be regarded as one point.)

As the beam scans a horizontal line, its intensity changes 435 times, producing darker or lighter dots on the screen; persistence of vision causes the viewer to perceive this as a whole picture instead of a series of dots.

The stroboscopic effect comes from the screen being alternately illuminated and dark.

127.

The phenomenon results from the interplay of electromagnetic waves of different wavelengths and several layers of ionized air known collectively as the ionosphere. The earth's atmosphere is continuously flooded by photons (light particles) from the sun, and cosmic rays. The highly energetic photons, primarily those belonging to the extreme ultraviolet or X-ray portion of the spectrum, have enough energy to split off an electron from neutral atoms in the upper atmosphere and leave positively charged ions. The cosmic rays carry much more energy than photons, but are much, much fewer in number, so their effect is small.

The several ionized layers are (1) the D layer, which disappears at night; (2) the E layer, which almost disappears at night; and (3) the F_1 and F_2 layers, which diffuse into each other during daylight hours. At night the F_1 and F_2 layers merge completely, mostly due to the disappearance of the F_1 layer. The resulting F layer becomes an excellent reflector of shortwaves.

How do these layers affect the propagation of electromagnetic waves? The low-frequency waves up to 500 kHz cannot penetrate very far into ionized air, and are reflected by the D layer. Medium-frequency waves, including AM radio broadcasts, are not bent as easily. They are only reflected by the E layer. High-frequency waves, that is, short-waves, get as far as the F layer before they are bent back toward the earth. Waves of very high frequency and above, including television and radar, carry so much energy that they escape into outer space. As a result, TV and radar are limited to line-of-sight distances, generally no more than 40 mi.

128.

In some areas households used to be supplied with 110 volt direct current. A higher voltage was considered unsafe. Later, such fears came to be regarded as exaggerated (see Answer 121), in most countries.

Higher voltages economize power distribution.

Power supplied by electric current equals voltage times current. If the voltage is doubled, the same power is delivered with half the current, which means conductors of half their former cross section, resulting in great saving of copper and other rapidly vanishing metals. A halved current flowing through a conductor of half its former cross section encounters unchanged resistance.

Efficiency is also gained in terms of I^2R (heat) losses. Assume an initial current of $I = 4$ amperes and a line resistance of 4 ohms. The I^2R loss $= 16 \times 4 = 64$ watts (lost in the form of heating the copper). Now double the voltage. The current is cut in half; $I = 2$ amperes. The cross-sectional area of the copper is cut in half, doubling the resistance to 8 ohms. So the loss becomes $4 \times 8 = 32$ watts.

129.

No, the principle of the conservation of energy has not been violated. As the capacitor C_2 was being charged, part of the energy went to heat the connecting wires and part was radiated away in the form of electromagnetic waves.

When $C_2 = C_1$ the energy loss is always 50 percent, no matter what the resistances of the connecting wires. (They have not even entered the calculation.)

130.

The difference exemplifies one of the most interesting things about magnetic (and electric) fields. The magnetic field of the earth is almost uniform across the length of the needle, that is, its intensity remains the same in direction and magnitude, because the distance between the ends of the needle is relatively short compared to the distance between the magnetic poles of the earth. Thus the magnetic force on one end of the needle is practically canceled by the force on the other end. The resulting couple can only twist the needle around, but as soon as the needle is aligned with the direction of the magnetic force, the resultant force practically disappears.

By contrast, the field of a magnet is highly non-uniform in its vicinity. Therefore, the forces acting on the poles of the needle are different, giving a non-zero resultant.

131.

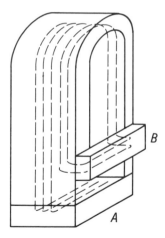

When bar *B* is placed on the magnet, some of the magnet's lines of force are "short-circuited" through bar *B*, as illustrated. As a result, the number of lines of force passing through bar *A* is sharply reduced.

This weakens the attractive force between the magnet and bar A, which drops off.

132.

It is possible to put together the sphere as described, but the sphere will be observed to have no magnetic properties. It becomes demagnetized as soon as it is put together. This can be seen on theoretical grounds: the sphere is perfectly symmetric. Therefore, through every point of the sphere pass equal numbers of oppositely directed magnetic lines of force, and the net effect at each point is zero.

10. Light and Vision

133.

Glass. The clue is the narrow space between the bottom of the can and its reflection. The mirror coating covers the bottom of the glass plate in most glass mirrors.

134.

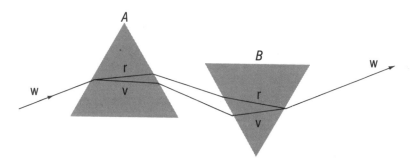

No. The first diagram illustrates this common fallacy, perpetuated to this day in many texts. The second diagram is correct.

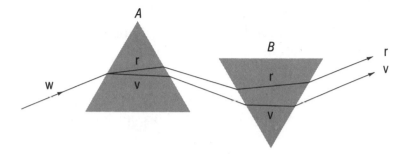

The paths of the red and violet rays shown in prism *B* must be parallel to the corresponding rays in *A*. It is just like what happens to a light ray passing through a glass slab with parallel faces. The only difference is that here the roles of glass and air are exchanged.

The third diagram shows the simplest arrangement of prisms (*A, B, C, D*) that will resolve and completely synthesize a beam of white light. Not two, not three, but four similar prisms are necessary.

The letters *r* and *v* on the diagrams refer to red and violet rays; the letter *w* to white light.

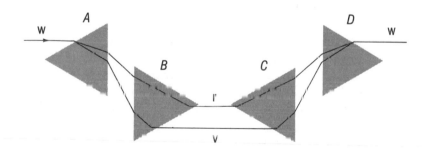

135.

No. To be perceived light must be absorbed by the retina. The retina of an invisible person would be transparent; it could not absorb light.

Moreover, his eye lens would be transparent. Its index of refraction would equal that of the air; therefore, it could not bend light to form an image on the retina.

136.

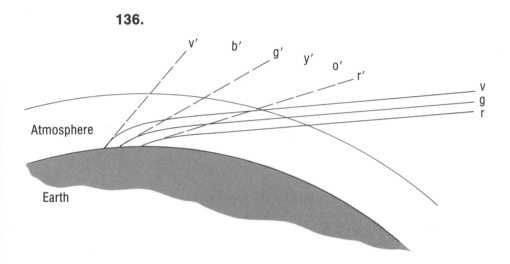

Light from the sun looks white to us, but passed through a prism it separates into all the colors of the rainbow. The earth's atmosphere behaves like a giant prism. It refracts (bends) the components of sunlight, the shorter waves (violet, blue) being bent more than the longer ones (red, orange, yellow). The amount of dispersion of white sunlight increases when sunlight passes through more air before reaching the observer, at sunset and sunrise. (The letters on the diagram refer to the colors of the various components.)

The diagram illustrates how the shorter waves are bent more sharply and appear to come from points higher in the sky than the longer waves. (The eye always assumes that a light ray comes from a point lying on the tangent to the path of the ray.)

Thus the sunlight spectrum has violet on top and red on the bottom. If a fair portion of the solar disk is visible above the horizon, the light rays from its various parts overlap and the spectrum cannot be seen; but as the sun sets the colors of its spectrum should theoretically vanish one by one, the red rays first and the violet rays last. However, two other atmospheric effects must be taken into account: the absorption of light, due mostly to water vapor, oxygen, and ozone, which screens out mostly the orange and yellow light; the scattering of light, with the shorter wavelengths (violet and blue) mostly affected. The only relatively unscathed color is green, which is what reaches our eyes. At high altitudes, where the air is usually clearer, the shorter waves may still come through, and the flash can be blue or violet instead of green.

The flash lasts longer if the sun sinks relatively slowly—in winter at any one place (since the sun's apparent path makes the smallest angle with the horizon then); and at all times nearer the poles. At Hammerfest, Norway (latitude 79° N) the flash at midsummer may last 14 minutes, 7 minutes during sunset and another during sunrise—which follows immediately!

137.

The principal role of the astronomical telescope is to present more light to the retina by substituting the large area of the lens (or spherical mirror) for the tiny area of the human pupil. If a stream of photons (light particles) falls on the earth from a distant star, the number of photons per unit area is constant. Therefore, a large telescope of area A_1 collects A_1/A_2 times as many photons as a telescope of area A_2. In terms of their diameters d_1 and d_2, the large telescope will collect d_1^2 / d_2^2 times as much light as the small telescope. The Mount Palomar telescope ($d_1 = 200$ in.) collects 440,000 times as much light as the human pupil ($d_2 = 0.3$ in.):

$$\frac{d_1^2}{d_2^2} = \frac{(200)^2}{(0.3)^2} = 440,000$$

By definition, a first-magnitude star is 100 times as bright as a sixth-magnitude star. Then the ratio between successive magnitudes is $(100)^{1/5} =$ about 2.512. If a star of apparent magnitude m_1 has intensity I_1 and one of magnitude m_2 has intensity I_2, then

$$\frac{I_1}{I_2} = 2.512^{(m_2 - m_1)}$$

and

$$\log_{10}\frac{I_1}{I_2} = (m_2 - m_1)\log_{10} 2.512 = 0.4(m_2 - m_1)$$

Hence a telescope with 10 times the area of another will see stars 2.5 magnitudes as faint.

Another advantage of using a telescope is its higher resolving power, given by

$$2.52 \times 10^5 \frac{\lambda}{d}$$

where λ is wavelength of light and *d* is diameter of the telescope. However, the resolution is limited severely by the atmosphere (Answer 168), the smallest resolution possible being 0″.5 of arc, obtainable with a 30 cm (12 in.) telescope. Larger telescopes only give a gain in light-gathering power.

138.

Raindrops can only assume a perfectly spherical shape if they are subject only to intermolecular forces. The spherical shape minimizes the energy of surface tension.

But air drag distorts the surface of a drop so that it assumes the characteristic raindrop shape. The reasoning in the problem does not apply to such raindrops, since the conditions of internal reflection are not the same at every point. A light ray can undergo internal reflection at one point in such a drop and leave the drop at another point.

Moreover, no internal reflection is complete: some energy always penetrates to the outside.

139.

The reflection coefficient for light reflected from the surface of water decreases when the angle of incidence becomes smaller.

When looking straight down, you receive rays reflected at very small angles. The rays are reflected from the water surface near the horizon at larger angles, so fewer of them are absorbed.

140.

No. The color we see depends not on wavelength but on the frequency of vibrations in the light wave. This is what stimulates the optic nerve and makes us perceive a certain color.

The frequency does not change as light passes from one medium into another. The formula given in the problem can be rewritten as

$$\frac{v_1}{\lambda_1} = \frac{v_2}{\lambda_2}$$

which means the ratio v/λ, which is the frequency, remains constant.

141.

Smoke has no true color. Color perception by humans is determined not only by the frequency of light waves falling on the retina but by psychological considerations including prejudice and ignorance. Thus paper known to be white appears white whether viewed in direct sunlight or in the yellow light of a candle. On the other hand, a fabric of unknown color may appear greenish under an incandescent lamp and bluish in the light of the northern sky. The color of smoke, which may range in composition from the highly reflective particles of certain tars to the highly absorbent particles of carbon black, depends on how the particles reflect the light and on the past experience of the onlooker. Smoke particles scatter shorter light wavelengths (that is, violet and blue) to the sides. For this reason, white light, after passing through a cloud of smoke, looks reddish, and when observed from the side looks bluish gray.

142.

Yes. The reflected landscape is seen as if viewed from a point below the surface of water at a depth equal to the distance between the objective of the camera and the water.

143.

The focal length of the eye or any lens depends on the wavelength of light. Red rays are bent less than blue rays. As a result, the focus of the red rays is farther away from the lens, and slightly behind the retina. Consequently, their image on the retina is slightly larger.

144.

It reflects less light. Light penetrates into the water between sand grains, travels inside through multiple internal reflections from the water boundaries, and is absorbed in the interior.

145.

More light is always reflected from the walls, which are opaque, than from the windows, which are transparent (that is, they let light in).

146.

If you stand close to a slab of red-hot metal you receive heat mostly through electromagnetic radiation, with a maximum of radiated energy in the infrared region. Metals are excellent reflectors of infrared (and all other electromagnetic) waves, so a metal coating provides an effective shield against heat radiated by a red-hot metal.

11. Spaceship Earth

147.

The surface tension of oil becomes greater as its thickness decreases, even down to a millionth of a millimeter. As the oil spreads away from where it is poured it becomes more effective in opposing any motion that tends to increase surface area. It becomes like an elastic membrane pressing down on the sea.

148.

Seawater creates more foam than fresh water primarily because of its solutes, including dilute slimes of organic materials.

Foam consists of air bubbles separated from each other by a film liquid. Bubbles colliding in fresh water coalesce, but in salt water they bounce off each other. This is why sea bubbles last longer.

Most ocean bubbles are caused by wind waves, but they may be produced by rain or even snow. The bubbles that form along the seashore are very small, mostly less than ½ mm in diameter.

149.

Because of the Coriolis force, which, caused by the rotation of the earth, makes everything sidle to the right of its motion (in the northern hemisphere). The prevailing winds that drive the water onto the California coast are from the northwest; this means the Coriolis force transports water away from the shore toward the southwest. The resulting deficit is made up by cold water rising from depths of several hundred feet, and a cool strip of water forms along the coast. The cold California current flows down from the north and lowers the temperature of the coastal waters even more.

150.

The side near the wall is exposed to a smaller area of open sky than the side away from the wall, so the near side will cool less by radiation into the sky than the far side, causing water vapor to be more likely to precipitate out on the cold far side.

151.

First, the upper surfaces of blades and leaves radiate heat into the sky. Simultaneously, stored heat from the subsoil is conducted to the base of stems. However, there is an efficient insulating layer between top and bottom consisting of stems and air spaces, which prevents the subsoil heat from moving upward.

Second, plants transpire vapor, saturating the air around them. On a clear still night the temperature in the grass can easily fall below dew point, and since cold air can hold less water vapor than warm air, extra vapor will condense out of the air and settle on the grass.

152.

Antarctica is a continent. Land is a poor heat conserver, radiating heat away as soon as it gets it. (This is why winters are harsh deep inland.)

The Arctic ice is over an ocean, and water is known for its high heat capacity, taking a long time to heat but once it is warm, losing heat slowly. The Arctic stores summer heat and lives off its "savings" in the winter.

153.

Salt is hygroscopic, that is, it tends to take up moisture from the air. Water molecules stick to the surfaces of salt crystals, and in so doing form bridges between the neighboring crystals, making the salt sticky.

154.

Smudge pots break up temperature inversion. Ordinarily, air temperature decreases with altitude at the rate of 3.3° F for each 1,000 feet. However, certain conditions may cause the temperature to increase with altitude. This is called temperature inversion.

On cloudless winter nights the earth's surface gives off a great deal of heat by infrared radiation.

The surface temperature drops sharply, cooling the lower layers of air much more than the upper ones. By sunrise, frost may wipe out an entire orchard if nothing is done to prevent it.

On cloudy nights the infrared rays emanating from land are absorbed by water vapor and cloud droplets before they have a chance to escape into space, and are then radiated back to earth. Thus no heat is lost, and the temperature doesn't drop as much as on cloudless nights.

Similarly, smudge pots produce a dense pall of smoke which spreads over the surrounding area and reduces the loss of heat by radiation. In addition to slightly raising the temperature of the surface air, heat from the smudge pots creates convection currents, which break up the inversion layer (see Answer 92).

155.

We must assume the tramp is sober enough to remember where he came from. All he has to do is replant the signpost so that the correct arm points back where he came from. The remaining four hands will automatically point in the right directions.

156.

At the geographical north pole. Every direction from the north pole, including the direction of the magnetic pole, is south.

157.

Snow on bare soil is melted by heat flowing up from the subsoil. Grass has air spaces among the blades which impede the conduction of heat to the snow on top of the grass.

12.   *The Universe*

158.

To speed it up!

    Considering a circular orbit, for simplicity, centripetal force equals gravitational attraction of the earth:

$$\frac{mv^2}{R} = Gm\frac{M}{R^2}$$

or

$$v = \sqrt{\frac{GM}{R}}$$

where M is mass of the earth and R is radius of revolution. Thus speed increases with a decrease in radius. But air drag, consuming the kinetic energy of the satellite, does precisely that—it decreases the radius of the orbit.

159.

It is easiest to launch a satellite along the equator in the direction of the earth's rotation (west to east). The earth's rotational speed is used in addition to the speed imparted by the rocket motors. The earth's linear speed due to rotation is largest on the equator.

160.

The easiest way is to use plastic bags or other elastic containers, and squeeze the liquid out as we do with a tube of toothpaste.

Another way is to use the law of conservation of momentum, which is valid in weightless states too. The astronaut places the containers end to end with their openings next to each other. Then he moves the containers sharply in the direction opposite to pouring. The liquid acquires an equal momentum to that of the container, but in the opposite direction. Now, the mass of the liquid is greater than that of the containers, and its acceleration will therefore be less than the acceleration of the containers. In effect, the astronaut slides the first container off the liquid and the second container onto it.

161.

The earth moves fastest in winter and slowest in summer (in the northern hemisphere).

The earth's path around the sun is slightly elliptical, which means the distance between the earth and the sun is constantly changing. Paradoxically for the inhabitants of the northern hemisphere, the earth is closest to the sun in winter and farthest away in summer. By Kepler's second law, the area swept out by the earth's radius vector remains constant. To sweep out as large an area the earth must move faster when it is close to the sun.

162.

The water creeps up the inner wall of the jar and out over the edge, finally covering both inner and outer walls.

Water, whether weightless or not, wets glass. Freed of its own weight, the water can easily be pulled by the forces of adhesion between water and glass.

163.

The occupants of a car are thrown to the outside of the curve because the car's centripetal acceleration is greater than the occupants'. The car is accelerated toward the inside of the curve by the frictional force of the road on the tires, whereas the occupants are only accelerated toward the inside of the curve by the frictional force of the seats on their bodies, which is small unless they are strapped or seated against the side of the car that goes on the outside of the curve.

In contrast, the spaceship and its occupants receive centripetal acceleration solely from the gravitational attraction of the earth, the only known force that gives all bodies the same acceleration no matter what their mass. (For example, if you push something, the smaller its mass, the greater the acceleration.) If a mass m acquires an acceleration a as a result of gravitational attraction by a mass M, we have

$$ma = \frac{GMm}{r^2}$$

But m cancels out and the resulting expression is independent of m.

Since the earth's gravity field gives the same centripetal acceleration to spaceship and astronauts, they will feel no acceleration with respect to the spaceship; relative to it they are weightless.

164.

An artificial earth satellite can only be seen if it is above the horizon and the sun is illuminating it from below the horizon. When the sun is in the sky it shines too brightly to allow you to see the satellite.

165.

On the earth we heat water mostly by convection. Heated water on the bottom of the kettle (near the heat source), being lighter, is displaced upward by the cold water on top, which sinks, gets heated, and rises again. These convection currents mix warm and cold water effectively.

There are no convection currents in a condition of weightlessness, since nothing is lighter or heavier than anything else. The water on top is only heated by heat *conduction*—a slow process, in water.

166.

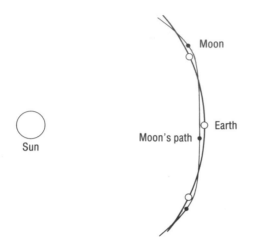

Yes. The moon's trajectory around the earth is always concave relative to the sun, that is, the moon always has a velocity component toward the sun. The path looks like a regular 13-sided polygon whose corners have been gently rounded.

167.

The sun gives the earth practically the same centripetal acceleration that it gives the moon. As we know (Answer 163), accelerations of bodies in a gravitational field are independent of their masses, so when we compare the moon and the earth, the only difference is their relative distance from the sun; but this is so small it can be neglected. Consequently, the paths of the earth and the moon around the sun are being curved at the same rate, so their mutual distance remains practically the same.

168.

Stars twinkle. Planets usually shine with a steady light except when near the horizon.

Stars are so far away that they appear as points even to the most powerful telescopes. The planets visible to the naked eye are so close that they can be seen as disks through a telescope. The apparent diameter of Venus is 10–65 seconds of arc, compared to 0.05 second of arc for the nearest stars. (To get a feeling for relative orders of magnitude, note that the average angular diameter of the moon is 31 *minutes* of arc.)

The light from a planet is made up of many individual light rays, not one ray (for all intents and purposes) as is true of a star. This means the light from a planet is relatively steady, because if a light ray from one point wavers because of atmospheric interference, say to the left, it is replaced by light rays from nearby points on the planet. The average intensity of the image on our retina remains the same.

169.

By coincidence, the orbit of the nearest possible satellite—one just grazing the atmosphere—is very nearly 90 minutes. Because 90 minutes is exactly one-sixteenth of a day, the earth rotating underneath, after 24 hours, will bring the satellite back to almost the same spot in the heavens.

INDEX